Denys Kataiev · Janusz Kacprzyk ·
Artur Zaporozhets

Neural Calibration of Coordinate Measuring Arms

Springer

Denys Kataiev
General Energy Institute of NAS of Ukraine
Kyiv, Ukraine

Artur Zaporozhets
General Energy Institute of NAS of Ukraine
Kyiv, Ukraine

Green Technology Research Center
Yuan Ze University
Taoyuan City, Taiwan

Janusz Kacprzyk
Systems Research Institute
Polish Academy of Sciences
Warsaw, Poland

ISSN 2198-4182 ISSN 2198-4190 (electronic)
Studies in Systems, Decision and Control
ISBN 978-3-032-32917-2 ISBN 978-3-032-32918-9 (eBook)
https://doi.org/10.1007/978-3-032-32918-9

This Springer imprint is published by the registered company Springer Nature Switzerland AG
The registered company address is: Gewerbestrasse 11, 6330 Cham, Switzerland

Preface

This book is devoted to the development of a software and information system aimed at improving the accuracy of geometric measurements of power equipment components using a coordinate measuring system that takes into account both kinematic and non-kinematic errors. Particular attention is paid to the application of artificial neural networks for compensating residual measurement errors that cannot be eliminated by conventional calibration methods.

Chapter 1 provides an analytical review of modern methods for error compensation in coordinate measuring arms. It has been established that the majority of existing approaches focus solely on the compensation of kinematic errors. A conceptual framework is proposed for combining classical modelling techniques with neural network-based methods to compensate for residual errors. The sources of destabilising factors are analysed, and the key research objectives of the dissertation are formulated, namely: to analyse the primary sources of kinematic and non-kinematic errors; to develop and improve the kinematic model of the information and measurement system; to propose methods for the compensation of kinematic errors; to design a methodology for constructing, training, and testing a neural network to compensate for residual non-kinematic errors; to implement a software-information system equipped with appropriate calibration and error compensation tools; and to carry out experimental investigations to assess the efficiency of the developed system in comparison with traditional approaches.

Chapter 2 describes the construction of the kinematic model of the coordinate measuring arm using Denavit–Hartenberg parameters. A combined calibration method based on three reference standards is proposed, which enables the unification of input parameters, accounts for nonlinear distortions, and ensures accurate reconstruction of spatial coordinates. A single-point correction model for residual kinematic errors is developed, employing polynomial approximation, which allows the residual kinematic errors to be reduced by a factor of four.

Chapter 3 presents the design of the architecture of a software-information system with an embedded neural network. The rationale for the selection of software tools, libraries, activation function types, and hyperparameters is provided. The use of a neural network architecture based on backpropagation has been justified. A structural diagram of the software-information system has been constructed, incorporating modules for calibration, filtering, and compensation utilising the neural network.

Chapter 4 is devoted to data preprocessing methods. A simulation approach for destabilising factors has been proposed by generating synthetic datasets based on a spherical reference object. An algorithm for data normalisation and partitioning into training, validation, and testing sets has been developed. An analysis of the model's sensitivity to data distribution and the influence of noise components has been conducted.

Chapter 5 presents the results of experimental investigations. The influence of the neural network architecture and training parameters on measurement accuracy and data processing speed has been examined. Based on an evaluation of 432 neural network architecture configurations, an optimal model was identified for compensating non-kinematic measurement errors. A comparative analysis of measurement accuracy before and after neural network-based compensation was conducted. It was demonstrated that the application of the neural network enabled a sixfold reduction in the non-kinematic component of the residual error.

The following new scientific results were obtained in this work:

1. A universal calibration method is proposed, which consists in the use of three calibration standards within one procedure, which makes it possible to take into account systematic errors in surface sensing, to assess the ability of the coordinate measuring arm to stably reproduce the same point in space and to determine the parameters of reproducibility of the linear movement of the coordinate measuring arm.
2. A method for compensating for the residual kinematic error using a single-point model of residual error correction was developed, which made it possible to reduce the kinematic component of the residual error by 4 times.
3. The theory of neural network training was further developed, which allowed to automatically take into account variable measurement conditions and compensate for the influence of destabilising factors, which made it possible to reduce the non-kinetic component of the residual error by 6 times.

The practical significance of the results obtained in the book is the development of a software and information system that allows to improve the accuracy of measurements of geometric parameters of power equipment parts in difficult production conditions. The structure of the information-measuring system based on the combination of the information-measuring system and the software-information system with an integrated neural network to improve the accuracy of measurements of geometric parameters of power equipment parts is proposed. A software and information system with an integrated neural network was developed, which allows for calibration, data processing, training of the neural network and automation of the processes of compensation for destabilising factors. 432 modifications of neural networks were created to analyse the current data of the process of measuring the geometric parameters of power equipment parts, which made it possible to take into account variable measurement conditions and compensate for the impact of destabilising factors.

The presented results have been implemented in the production processes of the following enterprises: AVA TECH LLC (Kyiv), in the development of a method for improving the measurement of geometric parameters of components and constructing three-dimensional representations (implementation certificate dated 01.02.2024); and ECO LUX LLC, in the development of a method for measurement control accounting for the influence of destabilising factors and the design of an algorithm for quality control procedures (implementation certificate dated 22.01.2024).

Kyiv, Ukraine
October 2025

Denys Kataiev
Janusz Kacprzyk
Artur Zaporozhets

Competing Interests The authors have no competing interests to declare that are relevant to the content of this manuscript.

Contents

Chapter 1
Modern Approaches to Reducing Kinematic and Non-kinematic Errors in Coordinate Measuring Arms

1.1 Introduction

An information and measurement system based on a coordinate measuring arm is one of the key tools in the production, maintenance, repair, prototyping and reverse engineering of power equipment. This device allows for system of three-dimensional measurements with high accuracy, which is especially important in critical industries such as the manufacture and maintenance of turbines, generators, and pumps for nuclear, thermal, and hydroelectric power plants. IMS is used at all stages of the equipment life cycle, providing quality control, part reproduction and the development of new solutions.

Sources of errors in information and measuring systems based on a coordinate measuring arm are usually divided into kinematic errors, such as link length errors, link twist errors, and initial zero position errors, and nonkinematic errors, such as link deformation errors, thermal errors, and encoder rotary shaft movement errors.

Most of the studies in this area were mainly devoted to the calibration of kinematic models and ignored the compensation of nonkinematic errors. Therefore, in order to compensate for non-kinematic errors, it is proposed to introduce a software and information system with an integrated neural network into the composition of the IMS based on the CMA [1–5].

The main idea is to compensate for the residual non-kinematic errors after compensating for kinematic errors classical methods with the help of a trained neural network, which is able to compensate for the influence of external destabilising factors in production conditions and thus increase the accuracy of measuring the geometric parameters of power equipment parts.

D. Kataiev et al., *Neural Calibration of Coordinate Measuring Arms*, Studies in Systems, Decision and Control 695, https://doi.org/10.1007/978-3-032-32918-9_1

1.2 Analytical Review of Modern Methods of Improving the Accuracy of Measurements Using CMAs

The main sources of CMA errors are kinematic errors (link length errors, link twist errors, initial zero position errors). The analysis of the current state of research shows that considerable attention is paid to the development of methods for calibration and compensation of such errors.

One of the most common models for describing the kinematics of a CMA is the Denavit-Hartenberg model (D-H) [6, 7], which is considered a classic model. However, its application has certain limitations in cases where the two axes of the joint are parallel or nearly parallel.

To solve this problem, a modified Denavit-Hartenberg model was proposed. In particular, Nguyen et al. used a full position measurement method to calibrate kinematic errors based on a circular trajectory [8]. Li et al. presented a six-parameter modified D-H model to describe the kinematic errors of the CMA, which includes axis parallelism, link length, and initial zero-position errors [9].

Sładek et al. proposed a virtual model of CMA based on an error compensation matrix to assess the accuracy of measurements in real time [10]. Zheng et al. created a model of the spatial distribution of CMA errors based on the theory of support vector machines, which allowed to obtain a measurement area with high accuracy without additional hardware costs [11].

Zheng has also proposed several measurement models to improve accuracy within the working area of the CMA. Dupuis et al. used the Gauss-Helmert method to calibrate kinematic errors through least-squares analysis [12].

However, most of these studies focus on the calibration of parametric models and are aimed at compensating for kinematic errors without taking into account non-kinematic errors. This is because kinematic errors are systematic and can be effectively modelled using classical methods.

To compensate for non-kinematic errors, Neumann et al. have developed a general approach to eliminate kinematic and non-kinematic errors in CMAs [13]. Gao proposed a general geometric error model to minimise the impact of non-kinematic errors in CMAs [14].

Huang presented a new approach to minimising volumetric errors based on the principles of Abbe and Bryan [15]. Santolaria developed a kinematic error model combined with a repeatability model method to improve the accuracy and reproducibility of measurements [16].

An analysis of current research confirms that the integration of neural networks into the error compensation process is a promising area. For example, Zhao proposed a model that uses a neural network to establish the relationship between spatial errors and the angles of the CMA joints [17]. Tian and Zhao developed a cooperative co-evolutionary neural network to determine the nonlinear relationship between connection variables and spatial errors of CMAs [18].

Thus, the compensation of nonkinematic errors remains less studied, as they arise under the influence of a large number of external factors and therefore their algebraic description is quite complex and, in some cases, impossible. In this regard, the use of methods based on the use of NNs is a promising direction.

Further research in this direction is focused on improved methods of compensation of kinematic errors of the CMA and the implementation of a software and information system on the basis of which it is proposed to combine classical methods of compensation of kinematic errors with further compensation of residual nonkinematic errors using NN.

1.3 Analysis of the Impact of Destabilising Factors and Sources of Errors

Information and measurement systems based on CMAs are high-tech devices that allow measuring the three-dimensional coordinates of product or part points without the use of bulky stationary structures (as in the case of coordinate measuring machines with gantry or cantilever configuration). Due to their mobility and flexibility, CMAs are widely used in the machine building, energy, automotive, aviation industries, as well as in scientific research involving the measurement of complex surfaces and non-standard parts. However, like any complex information and measurement system, the CMA is subject to various external factors that can significantly reduce the accuracy and repeatability of the results. The measurement process is influenced by a large number of destabilising factors that lead to kinematic and nonkinematic errors. The factors of occurrence of kinematic errors of IMS are given in Table 1.1.

When constructing a kinematic model of IMS, homogeneous matrix transformations or Denavit-Hartenberg parameters are used to describe the probe's position in 3D space [19]. Each joint defines a rotation around a certain axis, and each link is characterised by the length d_i, displacement a_i, and displacement angle α_i. In

Table 1.1 Factors of occurrence of kinematic errors in IMS

Type of kinematic error	Factor of occurrence
Geometric errors of the links	Inaccurate production of segment lengths; Deviation of the real axes of rotation from the nominal ones
Angular errors	Error in reading the angle θ_i; Installation error of the rotation sensor; Misalignment of the rotary joint
Parallelism and orthogonality errors	Deviation of the angles α_i from the nominal 90° (or other specified values); Non-perpendicularity of two consecutive axes of rotation
Nonlinearities and hysteresis in hinges	Backlash that occurs when changing the direction of movement; Plastic or elastic deformation of the joints

a simplified form, the full transformation matrix can be written as the product of successive transformations:

$$\mathrm{T_n} = \prod_{i=1}^{n} {}^{i-1}\mathrm{A_i}(\theta_i, \alpha_i, a_i, d_i), \tag{1.1}$$

where ${}^{i-1}A_i$ is the transformation matrix of the i-th link from the $(i-1)$ coordinate system to the i coordinate system.

To formalise the errors, adjusted transformation matrices are used, which take into account deviations from ideal geometric parameters. Suppose that for the *i-th* joint we have the following deviations $\Delta\theta_i$—angle measurement error; Δa_i—link length error; $\Delta\alpha_i$—deviation between the joint axes; Δd_i—translational displacement error.

Then the real transformation matrix $(i-1)$ *of* $\tilde{A}_i$ can be as follows:

$${}^{i-1}\tilde{A}_i = R_2(\theta_i + \theta_i)T_z(d_i + d_i)T_x(a_i + a_i)R_x(\alpha_i + \alpha_i), \tag{1.2}$$

where $R_z(\phi)$—rotation around the z-axis by an angle of; $\phi T_z(d)$—translational displacement along the z-axis by d); $T_x(a)$—translational displacement along the x-axis by; $aR_x(\phi)$—rotation around the x-axis by an angle of a.

Thus, the total transformation operator from the CMA base to the probe, taking into account the errors, can be represented as follows:

$$\tilde{\mathrm{T}}_\mathrm{n} = \prod_{i=1}^{n} {}^{i-1}\tilde{A}_i. \tag{1.3}$$

If we denote the ideal (nominal) transformation matrix by ${}^0T_n$, and the real (error-corrected) matrix by ${}^0\tilde{T}_n$, then the relative error of the final probe position can be estimated as

$$\Delta P = \Delta P_r - \Delta P_i, \tag{1.4}$$

where

$$\Delta P_r = \begin{bmatrix} x_r \\ y_r \\ z_r \end{bmatrix} = {}^0\tilde{T}_n \begin{bmatrix} 0 \\ 0 \\ 0 \\ 1 \end{bmatrix}, P_i = {}^{(0)}\, T_n \begin{bmatrix} 0 \\ 0 \\ 0 \\ 1 \end{bmatrix}. \tag{1.5}$$

In real conditions $\Delta\theta_i$, Δa_i, $\Delta\alpha_i$, Δd_i may not be constant, but depend on the load, temperature, condition of the joints, etc. Therefore, experimental methods of determining and approximating these values, as well as various calibration procedures are used [20].

As a result of the study of kinematic errors of IMS, it becomes clear that the sources of errors are quite diverse, and their impact can be significant in precision measurement work. Careful calibration, study of force and temperature effects, analysis of the variation of axial displacements and angular deviations allow to minimise the total deviations and ensure high accuracy of the system.

Thus, a detailed description and mathematical modelling of kinematic errors is the basis for further development of the IMS design, development of improved calibration methods, implementation of intelligent compensation algorithms, which generally improves the measurement accuracy and expands the possibilities of using IMS in industry and science. The overall measurement accuracy is also significantly affected by non-kinematic sources of errors (Table 1.2).

These errors are the result of physical, external, and software and algorithmic factors that are not directly reflected in the kinematic model, but can cause additional deviations from the true value of the measured coordinates [21].

Let's consider the effect of thermal deformation. The most common model of thermal expansion is the linear dependence of the link size on temperature.

Let's assume that the length of a certain element ℓ at a nominal temperature of T_0 changes when the temperature changes to ΔT_0 according to the formula:

$$\ell(T) = \ell_0[1 + \alpha(T - T_0)], \tag{1.6}$$

where ℓ_0 is the length at temperature, $T_0\alpha$ is the coefficient of linear thermal expansion of the material.

Since each link can have its own coefficient α, as well as uneven heating, the accuracy of measurements depends on the uniformity of the temperature field. Differential calculus principles and derivative error methods can be used to estimate the effect of thermal expansion on the probe endpoint coordinates. If $f(\ell_1, \ell_2, \ldots, \ell_n,)$ describes

Table 1.2 Factors of non-kinematic errors in IMS

Type of non-kinematic error	Factor of occurrence
Thermal deformation	Change the linear dimensions of the hand segments Temperature deformation of the measuring probe Uneven heating of individual links
Deformations due to loading	Elastic deformation of segments and joints The presence of backlash in the hinges Accumulation of errors due to material fatigue
Algorithmic and software errors	Digitisation and calculation errors Imperfection of data processing in the CBR controller Incorrect temperature compensation algorithms
Hardware errors of the encoder	Quantization error Optical and magnetic errors Contamination of the scale Temperature changes
Errors in the signal transmission process	Electromagnetic interference (EMI/EMC) Signal attenuation Distortion of analogue signals

the position of the probe (for example, using direct kinematic equations), then the error due to temperature effects ΔP_t will be equal to:

$$\Delta P_t \approx \sum_{i=1}^{n} \frac{\partial f}{\partial \ell_i} \Delta \ell_i. \tag{1.7}$$

Taking into account $\Delta \ell_i = \ell_0 \alpha_i (T - T_0)$, the total impact of thermal changes on spatial coordinates can be estimated.

Another important source of non-kinematic errors is elastic deformation (deflection). Let E be the Young's modulus of the material and I be the moment of inertia of the cross-section of the link. For the simplest case of a single-point load F applied to the free end of a rod of length L, the deflection w in the Euler-Bernoulli formula:

$$w = \frac{FL^3}{3EI}. \tag{1.8}$$

In real-world CMAs, the load may be more complex and distributed, but the principle remains the same: each link deflects under the weight of its own mass or under the action of the operator. This leads to a displacement of the endpoint by a certain amount ΔP_d. Vibrations acting on the measuring system can be described in general terms through harmonic vibrations:

$$x(t) = \mathrm{A}\sin(\omega t + \phi), \tag{1.9}$$

where A—amplitude of oscillations, ω—angular frequency, ϕ—initial phase.

During measurements, these vibrations cause fluctuations in the encoder readings and can make it difficult to stably fix the point of contact between the probe and the object. To reduce the impact of vibrations, damping materials, vibration isolation tables, or software filters that average the signal over time are used.

Algorithmic errors include: signal quantisation; digital filtering errors; errors in angle interpolation methods; errors in temperature and deformation compensation algorithms [22].

In a simplified form, the error due to algorithmic factors ΔP_a can be considered as a function of the quantisation step δ_q, bit depth n, sampling period T_s, etc.:

$$\Delta P_a = f\left(\delta_q, n, T_s, \ldots\right). \tag{1.10}$$

Non-kinematic errors can occur not only in the mechanical components of the coordinate measuring arm, but also in any sensor components, including encoders, as well as in the process of transmitting their signals to the controller or computing unit.

These errors belong to electronic and algorithmic sources and are also classified as "nonkinematic", as they are not caused by the geometry or kinematics of the links [23].

Hardware errors of the encoder include quantisation and errors, optical or magnetic errors (depending on the type of encoder), interpolation errors. Errors arising during signal transmission are caused by signal attenuation and distortion of the analogue signal.

Consider the quantisation of the encoder. If we have an incremental encoder with N pulses per revolution, the angular position of the rotor (or the joint of the CMA) after counting k pulses are given as:

$$\theta = \frac{360^\circ}{N} \cdot k. \tag{1.11}$$

The angular error associated with such a step can be within $\pm\frac{360^\circ}{2N}$ (half a step) as a simple approximation.

Consider the model of electrical noise [24]. Suppose that the signal from the encoder passes through a preamplifier and is digitised with rms noise σ_n. If we track the phase of the sinusoidal signal ϕ obtained from the cosine and sine channels, then (under the simplified hypothesis of small noise) we can write down an approximate relationship:

$$\Delta\phi \approx \frac{\sigma_n}{A}\cos(\phi) \quad \text{or} \quad \Delta\phi \approx \frac{\sigma_n}{A}\sin(\phi), \tag{1.12}$$

where A is the amplitude of the useful signal.

This shows that whenever the signal becomes close to zero (in the region of $\phi \approx 0$ or $\phi \approx \pi$), a small noise can cause significant phase jumps and, accordingly, deviations in pulse counting or interpolation.

The total angular error of the encoder $\Delta\varphi_e$ is the result of the superposition of independent error sources, so it must be determined by quadratic addition of the components, and it can be written in the form:

$$\Delta\theta_e = \sqrt{\Delta\theta_q^2 + \Delta\theta_t^2 + \Delta\theta_n^2}, \tag{1.13}$$

where $\Delta\theta_q$—static error of the sampling (quantisation) step, $\Delta\theta_t$—shift due to temperature changes (may be a slow drift), $\Delta\theta_n$—random signal distortion [25].

This model is often used to estimate the contribution of the encoder to the overall error of the CMA system. When the angle θ is further transformed into spatial coordinates (x, y, z) through a direct kinematic problem, each of the encoder errors $\Delta\theta_e$ in the joints gives an increase ΔP_e in the probe coordinate:

$$\Delta P_e \approx \sum_{i=1}^{n} \left(\frac{\partial f}{\partial \theta_i}\right) \Delta\theta_{e,i}, \tag{1.14}$$

where $f(\theta_1, \ldots, \theta_n)$ is the function describing the direct kinematic model, and n is the number of joints of the CMA.

For an integrated approach to estimating non-kinematic error, an uncertainty model based on the principles of the GUM (Guide to the Expression of Uncertainty in Measurement) is often used [26].

Then the total non-kinematic error ΔP_s can be considered as a vector function of a number of independent variables $x_1, x_2, \ldots, x_m$, each of which is associated with a specific source of error.

$$\Delta P_s = g(x_1, x_2, \ldots, x_m). \tag{1.15}$$

If each of these factors contributes to the standard deviation $u(x_i)$, then for small deviations, assuming mutual independence, the total standard deviation $u(\Delta P)$ can be found by the formula:

$$\sigma(\Delta P)^2 \approx \sum_{i=1}^{m} \left(\left. \frac{\partial f}{\partial \theta_i} \right|_{x_i = x_{i0}} \right)^2 \sigma(x_i)^2, \tag{1.16}$$

where $\left. \frac{\partial f}{\partial \theta_i} \right|_{x_i = x_{i0}}$ are partial derivatives of the error function for each source at the operating point x_{i0}. The final result can be represented as an expanded uncertainty $U = k \cdot u(\Delta P)$ with a coverage factor k, which is selected depending on the required probability level (for example, $k = 2$ approximately corresponds to a 95% confidence level) [27].

Non-kinematic errors play a significant role in the overall error balance of a coordinate measuring arm. Even a perfectly corrected kinematic model cannot provide high accuracy if the influence of external destabilising factors is ignored. Non-kinematic errors can make a significant contribution to the overall error balance: thermal deformations, backlash, encoder instability, signal distortion, which are complex, often nonlinear and unpredictable in nature [28].

Traditional approaches to modelling these impacts are not flexible enough, require an excessive number of assumptions and are sometimes impossible. In this regard, the task is to develop a software and information system with integrated NN, the implementation of which in the CMA will create an opportunity to implement a comprehensive method of compensating for non-kinematic errors, using NN capable of adaptive learning from real or simulated data.

1.4 Setting a Task for Development and Research

From the point of view of a practical problem, there is an urgent need to achieve high accuracy in the processes of measuring the geometric parameters of power equipment parts, despite the influence of external destabilising factors. At the same time, from a scientific point of view, it is clear that traditional mathematical and software methods are not always able to provide the required measurement accuracy under such complex influences [29]. This creates a contradiction: the industry requires methods that can be both highly intelligent (adapt to changing operating conditions) and easy to implement, but existing tools are either too highly specialised or do not fully take into account all destabilising factors.

The scientific and technical task can be formulated as the development and improvement of methods for improving the accuracy of measurements of geometric parameters of power equipment parts by developing and implementing a software and information system based on NN in the CMA, which will improve the accuracy of measurements in the presence of complex destabilising factors (Fig. 1.1).

To solve this problem, you need to do the following:

1. Analysis of existing approaches and systems:
   - Identify the advantages and disadvantages of traditional methods of controlling geometric parameters using IMS.
   - Evaluate the capabilities and limitations of common mathematical models and software tools used for calibration and error compensation.
2. Development and justification of the concept of a software and information system:

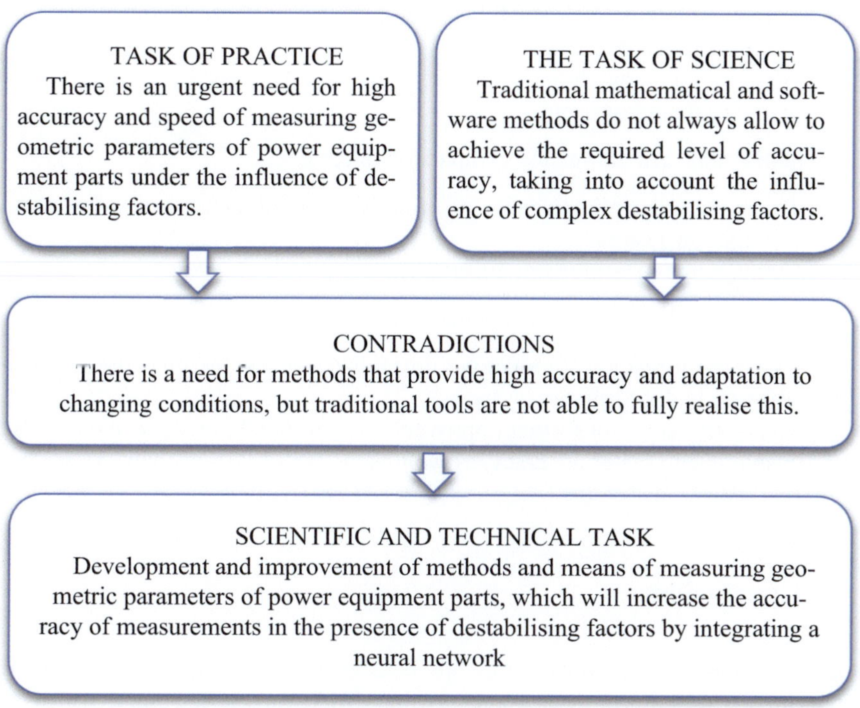

Fig. 1.1 Setting a scientific and technical task

- Build a kinematic model of the IMS to identify and minimise systematic errors.
- To propose the structure of the information and measurement system, where the main attention is focused on adaptive compensation of nonkinematic errors.

3. Modelling of non-kinematic errors and preparation of training data:
   - To simulate the generation of points on the surface of a reference body (e.g., a sphere), taking into account synthetically generated noise that simulates the effects of destabilising factors (temperature, hardware).
   - Create training and test samples for the initial training of the neural network, taking into account different noise configurations.
4. Developing machine learning methods to compensate for errors:
   - Justify the choice of NN architecture and optimal hyperparameters (number of layers, neurons, activation functions, regularisation methods, etc.)
   - Implement a suitable model (in Python or another platform) and train on the generated synthetic data.
5. Integration and research in a simulation environment:
   - Combine the kinematic model and the software and information system in a single software implementation.
   - Organise a computational experiment that simulates the operation of a IMS in a virtual environment.
6. Experimental verification and comparison of results:
   - To evaluate the effectiveness of error compensation using a software and information system.
   - Compare the measurement uncertainty.
7. Summarise and develop recommendations:
   - Based on the results of simulations, formulate conclusions about the developed software and information system.
   - To identify areas for further improvement of artificial intelligence methods to ensure even greater accuracy, as well as the possibility of their extension to real industrial systems.

1.5 Conclusions

In the first chapter of the thesis, a comprehensive analysis of modern methods for improving the accuracy of measurements using CMAs was carried out, as well as the main factors affecting the accuracy of such measurements were analysed. The analysis of literature sources has shown that traditional approaches are mainly focused on compensating for kinematic errors, while nonkinematic errors are either not taken into account at all or are considered in a highly simplified form.

It has been established that nonkinematic errors are characterised by a significant number of different destabilising factors that are complex, nonlinear in nature and vary with operating conditions and time. This makes it virtually impossible to create an effective analytical compensation model or compensation polynomial that would take into account the full range of effects of these factors.

In this context, the use of neural networks capable of adaptive learning and processing a large number of input parameters is particularly promising. The idea of combining classical kinematic approaches with neural network algorithms is proposed and substantiated, which will allow implementing a comprehensive method of error compensation. It has been established that the development of a software and information system, where the coordinate measuring arm will serve as a base for collecting spatial data, and the NN will recognise and correct the influence of destabilising factors, opens up new opportunities for automation and improving the accuracy of measurements of geometric parameters of power equipment parts.

Based on the analysis, the main tasks of the dissertation research are formulated: to analyse and formalise the main sources of kinematic and nonkinematic errors; to develop and improve the kinematic model of the information and measuring system; to propose methods of compensation of kinematic errors; to develop a methodology for constructing, training and testing neural networks to compensate for residual nonkinematic errors; to implement a software and information system with appropriate software and mathematical means of calibration and compensation.

References

1. Kvasnikov, V.P., Kataiev, D.A.: Methods of increasing the accuracy of measuring with the help of a coordinate-measuring hand. Cent. Ukr. Sci. Bull. Tech. Sci. **6**(37), 52–60 (2022). https://doi.org/10.32515/2664-262X.2022.6(37).1.52-60
2. Kvasnikov, V.P., Kataiev, D.A., Kvashuk, D.M.: Method of improving measurement reliability using a coordinate measuring arm in real-time mode. Prob. Inform. Manage. **4**(72) (2023). https://doi.org/10.18372/2073-4751.72.17459
3. Kataiev, D.A., Kvashuk, D.M., Dumbrava, S.M.: Accuracy of measurement of mechanical quantities using the theory of fuzzy sets. Prob. Inform. Manage. **1**(73) (2023). https://doi.org/10.18372/2073-4751.73.17638
4. Zaporozhets, A.., Kataiev, D.: Method of compensating for instrumental uncertainty in measurements using a coordinate measuring arm. Syst. Res. Energy (2024). https://doi.org/10.15407/srenergy2024.01.045
5. Shkvarnytska, T.Y., Yehorov, S.V., Kataiev, D.A., Kataieva, M.O., Molchanova, K.V.: Synthesis of analog matched filters for signals of different durations. Sci. Technol. Today. **4**(32), 858–870 (2024)
6. Denavit, J., Hartenberg, R.: A kinematic notation for lower-pair mechanisms based on matrices. J. Appl. Mech. **22**(2), 215–221. https://doi.org/10.1115/1.4011045
7. Denavit, J., Hartenberg R.: Kinematic synthesis of linkages. In: McGraw-Hill series in mechanical engineering. McGraw-Hill, New York, p. 435. Archived from the original on 28 Sept 2013. Retrieved 13 Jan 2012

8. Nguyen, H.-N., Zhou, J., Kang, H.-J.: A new full pose measurement method for robot calibration. Sensors **13**(7), 9132–9148 (2013). https://doi.org/10.3390/s130709132
9. Zheng, D., Xiao, Z., Xia, X.: Multiple measurement models of articulated arm coordinate measuring machines. Chin. J. Mech. Eng. (English Edition). **28**(5) (2015). https://doi.org/10.3901/cjme.2015.0716.095
10. Sładek, J., Ostrowska, K., Gąska, A.: Modeling and identification of errors of coordinate measuring arms with the use of a metrological model. Measurement **46**(9), 2785–2796 (2013). https://doi.org/10.1016/j.measurement.2012.08.026
11. Zheng, Y., Huang, W., Hu, P.-H., Liu, W.-W., Ye, B.: Design and validation of a self-driven joint model for articulated arm coordinate measuring machines. Appl. Sci. **9**(15), 31–51 (2019). https://doi.org/10.3390/app9153151
12. Romdhani, F., et al.: Methodology for the assessment of measuring uncertainties of articulated arm coordinate measuring machines. Meas. Sci. Technol. **25**(12) (2014). https://doi.org/10.1088/0957-0233/25/12/125008
13. Meng, Y., Zhuang, H.: Autonomous robot calibration using vision technology. Robot. Comput. Integr. Manuf. **23**(4) (2007). https://doi.org/10.1016/j.rcim.2006.05.002
14. Gao, G., Zhao, J., Na, J.: Decoupling of kinematic parameter identification for articulated arm coordinate measuring machines. IEEE Access **6**, 50433–50442 (2018). https://doi.org/10.1109/ACCESS.2018.2868497
15. Dolinsky, J.U., Jenkinson, I.D., Colquhoun, G.J.: Application of genetic programming to the calibration of industrial robots. Comput. Ind. **58**(3) (2007). https://doi.org/10.1016/j.compind.2006.06.003
16. Santolaria, J., Aguilar, J.J., Yagüe, J.A.: Kinematic parameter estimation technique for calibration and repeatability improvement of articulated arm coordinate measuring machines. Precis. Eng. **32**, 251–268 (2008). https://doi.org/10.1016/J.PRECISIONENG.2007.09.002
17. Zhao, H., Yu, L., Cao, J., Jia, H., Jiang, Y.: Improving the measuring length accuracy of articulated arm coordinate measuring machine using artificial neural network. Rev. Sci. Instrum. **91**(12), 106–125 (2020). https://doi.org/10.1063/5.0028365
18. Santolaria, J., Aguilar, J.-J., Guillomía, D., Cajal, C.: A crenellated-target-based calibration method for laser triangulation sensors integration in articulated measurement arms. Robot. Comput. Integr. Manuf. **27**(2), 282–291 (2011). https://doi.org/10.1016/j.rcim.2010.07.008
19. Santolaria, J., Majarena, A.C., Samper, D., Brau, A., Velázquez, J.: Articulated arm coordinate measuring machine calibration by laser tracker multilateration. Sci. World J. **11**, 681853 (2014). https://doi.org/10.1155/2014/681853
20. Gao, G., Lu, J., Yang, H.: Study on the structural parameter coupling of articulated arm coordinate measuring machines. Telkomnika **11**(5), 2454–2459 (2013). https://doi.org/10.11591/telkomnika.v11i5.2474
21. Cheng, W., Fei, Y., Yu, L., Yang, R.: Probe parameters calibration for articulated arm coordinate measuring machine. In: Proceedings of the 6th International Symposium on Precision Engineering Measurements and Instrumentation. Proceedings of SPIE, Hangzhou, China (2010). https://doi.org/10.1117/12.885303
22. Zhu, L.Q., Li, W.X., Pan, Z.K., Guo, Y.K., Chen, Q.S.: Coordinate measuring method with two operation modes based on the adjustable articulated arms. Opt. Eng. **53**(12) (2014). https://doi.org/10.1117/1.oe.53.12.122407
23. Cheng, Y., Wang, Z., Chen, X., Li, Y., Li, H., Li, H., Wang, H.: Evaluation and optimization of task-oriented measurement uncertainty for coordinate measuring machines based on geometrical product specifications. Appl. Sci. **9**(1), 6–15 (2016). https://doi.org/10.3390/app9010006
24. Hamana, H., Tominaga, M., Ozaki, M., Furutani, R.: Calibration of articulated arm coordinate measuring machine considering measuring posture. Int. J. Autom. Technol. **5**, 109–114 (2011). https://doi.org/10.20965/ijat.2011.p0109
25. Brau, A., Valenzuela, M., Santolaria, J., Aguilar, J.J.: Evaluation of different probing systems used in articulated arm coordinate measuring machines. Metrol. Measurem. Syst. **21**(2), 233–246 (2014). https://doi.org/10.2478/mms-2014-0020

26. Sładek, J., Ostrowska, K., Sokal, G., Kmita, A.: Calibration of coordinate measuring arms. Acta Mech. Autom. **2**, 53–58 (2007)
27. Rongsheng, L., Wanhong, L., Dabao, L.: Angular error compensation for laser tracker. Opt. Precis. Eng. **22**(9), 2299–2305 (2014). https://doi.org/10.3788/OPE.20142209.2299
28. Wang, Y., Xue, Z., Huang, Y., Wang, X.: Study on self-calibration angle encoder using simulation method. In: Seventh International Symposium on Precision Mechanical Measurements, vol. 9903 (2016). https://doi.org/10.1117/12.2217638
29. Zaporozhets, A., Sverdlova, A.: Predicting anomaly conditions of energy equipment using neural networks. E3S Web Conf. **280**, 10 (2021). https://doi.org/10.1051/e3sconf/202128009005

Chapter 2
Methodology for Reducing Kinematic and Dynamic Errors in Coordinate Measuring Arms

2.1 Introduction

In industrial environments, versatile and highly accurate information and measurement systems are becoming increasingly important, capable of providing stable results even under difficult operating conditions. Coordinate measuring arm-based IMSs have become one of the most convenient tools due to their compactness and the ability to be deployed quickly without the need for additional stationary bases. At the same time, the accuracy of IMS data can be significantly reduced due to unaccounted-for deviations in geometry, assembly and installation errors. These factors lead to discrepancies between the actual and theoretical parameters of the system, which directly affects the measurement results [1].

The second section of the paper is devoted to the compensation of kinematic errors, the creation of a kinematic model of an information and measurement system (IMS), as well as the development of methods and measures to improve its spatial accuracy. The chapter discusses in detail the use of the Denavit-Hartenberg parameters for a formal description of the connections and links of the CMA. An in-depth analysis of the impact of deviations in geometric parameters, such as link lengths, angles between links, and initial encoder displacements, on the accuracy of determining the final probe position is carried out. Calibration procedures using various reference objects are proposed, as well as a method for constructing polynomial models to compensate for kinematic errors. The implementation of this approach allows for effective compensation of kinematic errors in a real production environment, which contributes to the accuracy, repeatability, and reproducibility of measurement results [2–5].

D. Kataiev et al., *Neural Calibration of Coordinate Measuring Arms*, Studies in Systems, Decision and Control 695, https://doi.org/10.1007/978-3-032-32918-9_2

2.2 Application of the Kinematic Model of CMA

A kinematic model based on the Denavit-Hartenberg parameters was used to determine the geometric coordinates of the elements of the information and measurement system based on the CMA (Fig. 2.1). The use of the D-H model allows for a formal description of the geometry and relative position of the CMA links and joints, which greatly simplifies the mathematical description of the system and allows for deviations in the lengths of segments, angles between links, and shifts in the initial positions of encoders. This makes it possible to effectively compensate for kinematic errors [6].

On the basis of the constructed kinematic model of the CMA, the following coordinate system transformation matrices are formulated:

$$
{}^{i-1}A_i = \begin{bmatrix} \cos\theta_i & -\cos\alpha_i \sin\theta_i & \sin\alpha_i \sin\theta_i & l_i \cos\theta_i \\ \sin\theta_i & \cos\alpha_i \sin\theta_i & -\sin\alpha_i \cos\theta_i & l_i \sin\theta_i \\ 0 & \sin\alpha_i & \cos\alpha_i & \lambda_i \\ 0 & 0 & 0 & 1 \end{bmatrix}, \tag{2.1}
$$

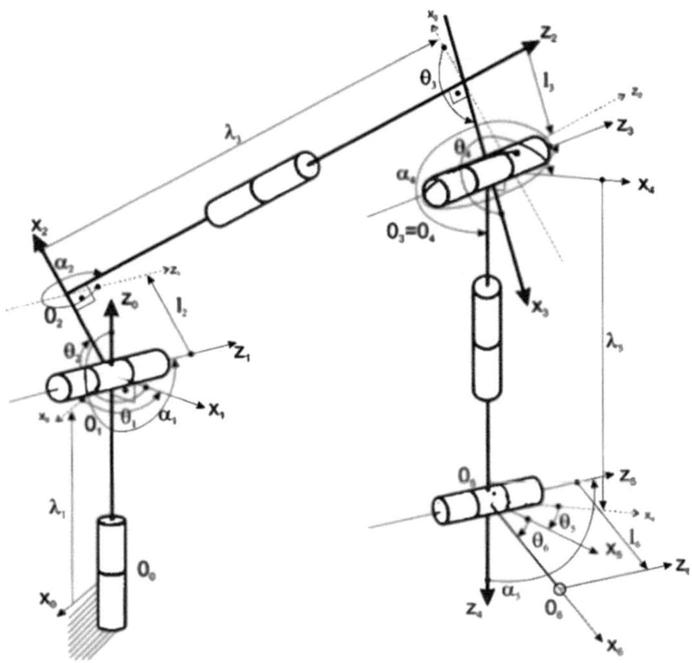

Fig. 2.1 Kinematic model of the CMA

where α_i are the angles between the z-axis, θ_i are the configuration coordinates reflected by the angles between the x-axis, l_i is the length calculated along the x-axis, λ_i is the length calculated along the z-axis.

The matrix A contains data describing the position and orientation of each respective CMA connection and can be represented as:

$$^{i-1}A_i = \begin{bmatrix} B & p \\ 0\,0\,0 & 1 \end{bmatrix}; \tag{2.2}$$

$$B_i = [l_i \ m_i \ n_i] = \begin{bmatrix} l_{ix} & m_{ix} & n_{ix} \\ l_{iy} & m_{iy} & n_{iy} \\ l_{iz} & m_{iz} & n_{iz} \end{bmatrix}; \tag{2.3}$$

$$P_i = \begin{bmatrix} p_{ix} \ p_{iy} \ p_{iz} \end{bmatrix}^T, \tag{2.4}$$

where B is the orientation matrix of each respective joint, P is the position matrix of each respective joint p_{ix}, p_{iy}, p_{iz} are the position coordinates of each respective joint $l_{ix}, l_{iy}, l_{iz}, m_{ix}, m_{iy}, m_{iz}, n_{ix}, n_{iy}, n_{iz}$ are the vector coordinates describing the rotation of all respective parts.

The problem of direct kinematics is a standard procedure in the study of mechanical manipulation. This is a static geometric problem that converts the configuration coordinates of a device into Cartesian coordinates [7].

The matrix of position and orientation of system n relative to the system associated with the i-junction can be defined as the product of the following transformations:

$$T_{n,i} = A_{i+1}A_{i+2}A_{i+3}\ldots A_n, \tag{2.5}$$

where A is the matrix that defines the position and orientation of each corresponding segment T is the transformation matrix of A.

Determining the accuracy of the geometric elements of a coordinate measuring arm is a complex and multistage task. The first step in solving this problem is to formulate and solve the problem of direct kinematics, which allows determining the coordinates and orientation of the measuring probe. To build the equations of direct kinematics, the parameters provided in the manufacturer's technical documentation are usually used. However, it is important to keep in mind that these parameters are only approximate and do not always correspond to the actual design features of a particular CMA. The differences between the parameters specified by the manufacturer and the actual values can be significant, which makes it necessary to determine the actual geometric parameters of the information and measurement system.

These real-world parameters include segment lengths, eccentricities, angles between hinge axes, and the so-called "zero-offsets", which characterise the difference between the actual encoder readings and their nominal initial values. These parameters are independent of the configuration of the CMA and remain constant in all its operating positions.

To accurately determine these parameters, it is necessary to conduct a series of measurements throughout the entire working volume of the CMA. Each such series should include at least 22 measurements, which will provide sufficient data to form a system of equations and solve it with respect to 22 unknown parameters.

First, it is necessary to set the coordinates of the configuration θ_i, as well as the Cartesian coordinates x_m, y_m, z_m. After obtaining the measurement data, these values of should be substituted into the previously formed system of equations, which will allow determining the exact geometric parameters of the CMA.

Thus, at this stage of calculations, the parameters that are usually considered constant when solving a direct kinematics problem are treated as unknown variables. At the same time, the data on the position and orientation of the finite element of the CMA are taken as known constant values read from the software of the coordinate measuring arm. These quantities include the coordinates x_m, y_m, z_m. measured by the arm at each point, the unit vectors A_m, B_m, Z_m that determine the orientation of the probe, and the angles θ_i read by the encoders at the time of each individual measurement. The algorithm for determining the geometric parameters of the kinematic model is shown in Fig. 2.2.

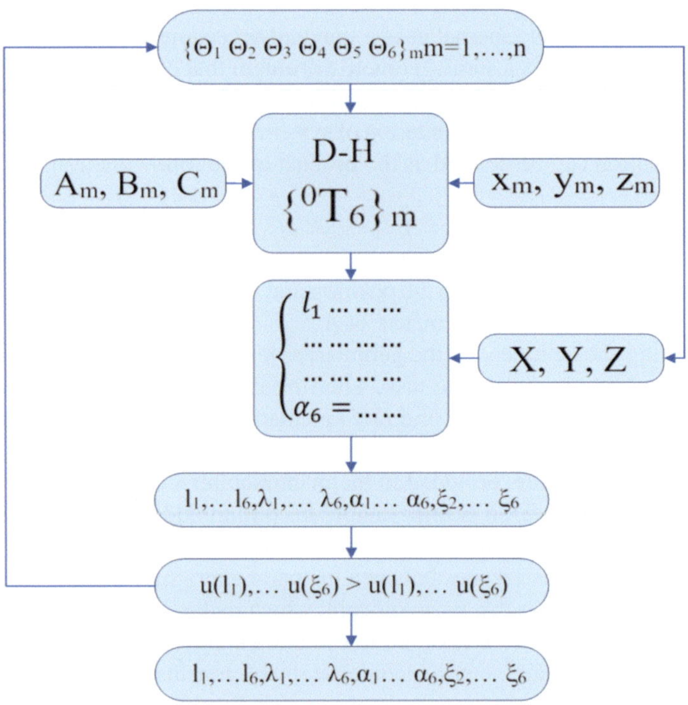

Fig. 2.2 Algorithm for calculating the geometric accuracy of the CMA

In Fig. 2.2 the following notations are given: Θ_m—measurements from angular measuring systems; A_m, B_m, C_m—units of the probe orientation vector; x_m, y_m, z_m—coordinates obtained from the software of the measuring arm; *X, Y, Z*—coordinates obtained from the direct kinematics problem; l_m—distance from the z_{m-1} axis to the z_m axis measured along; $x_m\alpha_m$—angle between the z_{m-1} axis and the z_m axis measured around; $x_m\lambda_m$—distance from the x_{m-1} axis to the x_m axis measured along; $z_{m-1}\xi_m$—"zero displacements" of the encoders.

Discrepancies in the values of the CMA parameters mainly arise due to errors made at the stages of production and installation of its elements. For example, during the installation of an encoder, it may be unintentionally displaced, which leads to errors in the "zero-offset". Another common source of errors is deviation from the perpendicularity of the hinge axes, which is reflected in the differences in the values of the parameters α. The presence of even minor shifts in the encoders can cause significant discrepancies between the coordinates of points determined by a mathematical model built using real parameters and those given in the manufacturer's documentation [8]. Therefore, after establishing the real parameters, it is necessary to carry out the calibration procedure of the CMA.

2.3 Development of a Combined Calibration Method

Let's look at modern calibration methods, their advantages and disadvantages. One of the methods is laser tracking, which provides high measurement accuracy and allows you to detect even minor displacements between links. However, a significant limitation of laser tracking is the high cost of the equipment, as well as sensitivity to negative external factors such as vibrations, dust, glare, and temperature fluctuations.

Interferometric methods are used to achieve high accuracy, as they allow for extremely sensitive detection of even the smallest changes in distances or angles during the movement of the CMA. However, this high sensitivity is also a major disadvantage, as uncontrolled temperature changes, vibrations, and airflows significantly affect the results in a production environment. In addition, the setup of interferometric equipment is quite complex and can lead to long work stoppages, and the need for precise synchronisation of the hand movement with the instruments creates additional requirements for software and electronics [9].

The self-calibration method is based on the measurement of a number of geometric standards and is used to determine the systematic errors of the CMA in different orientations. By comparing the obtained coordinates with the known nominal dimensions of the standards, compensation factors are determined, which significantly improve the accuracy in areas of frequent use of the device. At the same time, the quality and stability of the standards are critical, since any shape or dimensional inaccuracies can lead to distortion of the calibration results.

Since none of these approaches is universal and each has significant limitations, we propose a method for calibrating CMAs that combines several tests and takes into account the influence of different sources of errors within a single procedure.

This combined method is the most effective, since traditional methods do not provide stable accuracy over the entire operating range of the CMA. Combining the effective diameter test, single-point measurement test and volumetric measurement test into a single procedure allows compensating for the kinematic errors of the CMA, as well as increasing the repeatability of point measurements and the reproducibility of measurements over long distances.

If the geometry of the reference is known, it is possible to detect and compensate for discrepancies between the calculated and actual coordinates. To evaluate the accuracy of CMA measurements at different distances and in different orientations, appropriate calibration instruments are required.

The calibration sphere allows to determine the deviation of the effective diameter, the three-ball triangular socket is used to assess the repeatability of point measurements, and the ball rod allows to assess the accuracy of volumetric measurements when changing the orientation or angles of the CMA segments [10].

Each of these tests focuses on a specific source of error. The effective diameter test detects the deviations that accumulate when scanning spherical surfaces with minimal link movements and allows you to see possible hysteresis in the system. The single-point measurement test is focused on assessing repeatability, analysing the stability of the results with minimal probe displacements or even after moving the arm between separate measurements of the same geometric point. The volumetric measurement test, using a ball rod, provides information on linear distortions within a large working space and in different CMA orientations, simulating real production conditions in which CMAs are used to control complex geometric shapes.

After analysing three independent sources of error data, it becomes possible to use a residual correction method based on polynomial models. The essence of this method is to build a polynomial that approximates the real kinematics of the CMA based on information about the angles of the joints read from the encoders and the detected systematic displacements, which allows calculating correction factors and automatically adjusting the coordinates of the measured points in accordance with the configuration of the CMA.

The residual kinematic errors are mainly manifested in places of the largest deflections or deformations caused by the weight of the links and backlash in the joints. Polynomial corrections significantly reduce the impact of such errors. At the same time, polynomial models have certain disadvantages, including limited generalisation capabilities when extrapolating beyond the calibration range, sensitivity to the quality of input data, and high complexity as the polynomial degree increases, which can lead to model overtraining and deterioration of accuracy in real-world conditions.

The proposed combined calibration method not only improves accuracy, but also allows timely detection of faults that may manifest themselves as sudden multidirectional deviations from nominal values. From the perspective of energy or machine-building production, such a procedure can reduce the risk of defects in the production of large, complex geometry components, save resources by reducing rejects, and minimise the number of repeated measurements and additional adjustments [11].

The CMA calibration procedure is divided into three different tests: effective diameter test, single point measurement test and volumetric measurement test. For

Fig. 2.3 Calibration sphere

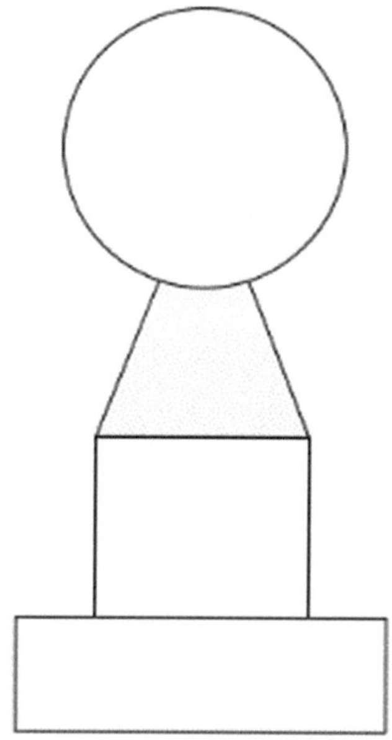

this reason, several calibration devices are required to perform the calibration, such as a calibration sphere, a three-ball triangular socket and a ball bar.

A calibration sphere is used for the effective diameter test and the single-point measurement test (Fig. 2.3).

The effective diameter test consists of nine-point measurements of the surface of the calibration sphere (Fig. 2.4).

Using this information, the IMS software defines the sphere and calculates its diameter. The measurement process should be performed with the minimum of articulation of the CMA. The difference between the measured and calibrated diameters of the sphere is reported as the diameter deviation. The effective diameter test value is reported as the maximum deviation, regardless of sign. The results can reveal various problems, such as hysteresis of the CMA or probe, or incorrect probe calibration [12].

For the single-point test, a three-ball triangular socket consisting of three steel spheres fixed to the bottom of the vessel is used. The three-ball triangular socket is used for the single-point measurement test with solid probes and, thanks to its three contact zones, provides a stable fixation of the CMA probe tip (Fig. 2.5).

The single-point measurement test is aimed at assessing the ability of the CMA to provide similar point coordinate values. This test includes aspects such as repeatability and reproducibility. If the CMA includes a solid probe, a three-ball triangular socket is used, but if a remote probe is used, a calibration sphere is used. The single-point test procedure consists of placing the three-ball triangular socket, or calibration sphere, in three different locations in the CMA working volume (Figs. 2.6 and 2.7).

The horizontal plane of the base is taken as a reference in the CMA measurement space, and several positions of the CMA segments are selected within 20–80% of the

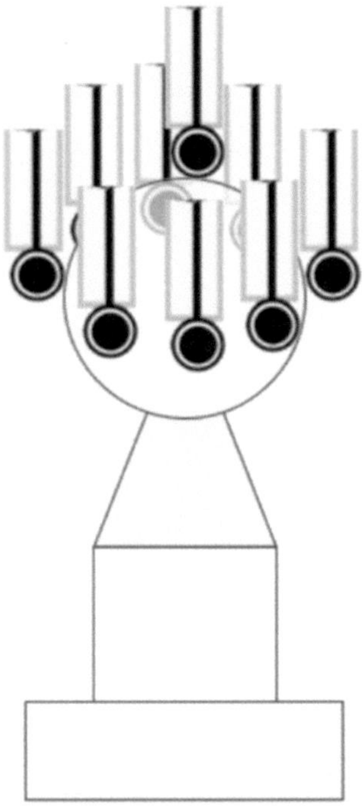

Fig. 2.4 Effective diameter test

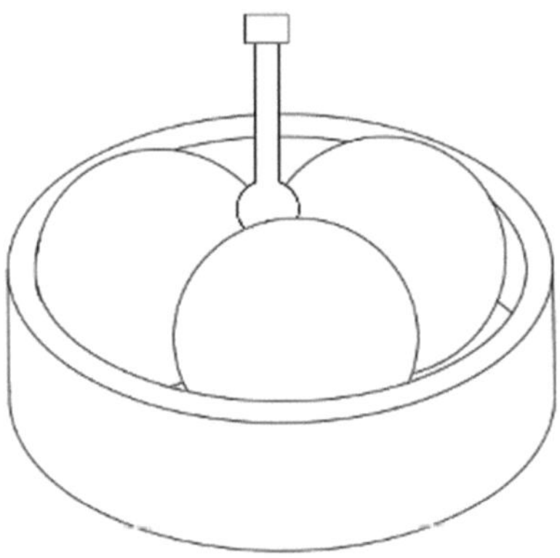

Fig. 2.5 Three-ball triangular socket

radial direction of the CMA measurement. In these places, using the method shown in Figs. 2.5, 2.6, data is collected from one point.

The result of the single-point test is the maximum deviation of the points from the mean and the double standard deviation of the point location [13].

A ball bar is used for the volumetric measurement test. The maximum calibrated length of the ball bar is 2400 mm. It is made of composite carbon fibres and consists of 13 balls, with a distance of 200 mm between two consecutive balls. Thus, the ball rod has a minimum calibration length of 200 mm and a maximum calibration length

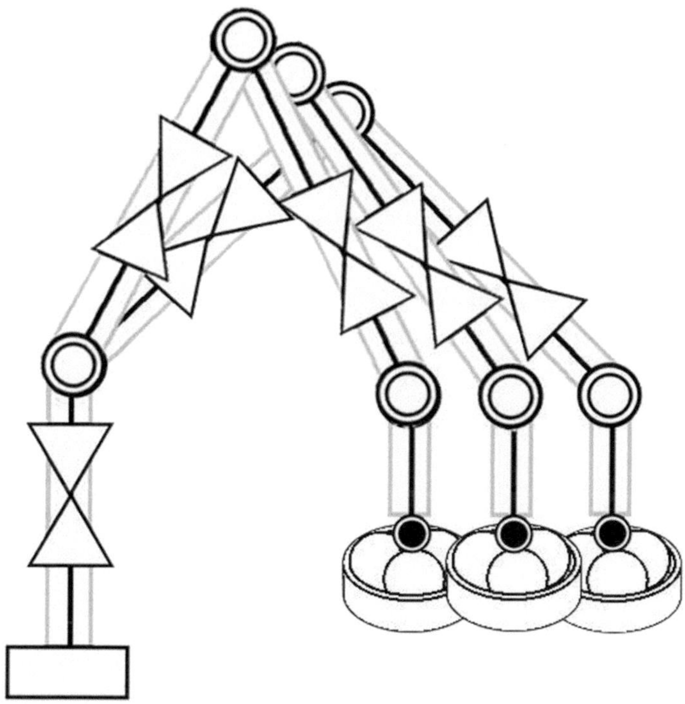

Fig. 2.6 Definition of a single point position

of 2400 mm, which allows for calibration with a sphere diameter of up to 4 m. The ball rod is fixed on a precision rotary table to position it horizontally and vertically, as well as at an angle of 45 (Fig. 2.8).

Prior to the calibration procedure, the CMA must be installed to the end user's specifications and ensure that the mounting is stiff and orientated appropriately. In addition, appropriate environmental, power and vibration conditions must be maintained throughout the process.

The calibration procedure consists of an effective diameter test, a single-point measurement test, and a volumetric measurement test [4].

The volumetric measurement test defines the linear displacement accuracy and volumetric accuracy aspects of the CMA as shown in Fig. 2.9.

The measurement process takes into account five measurement variables: ball rod length, working octants, ball rod inclination, distance from the CMA to the calibration sphere, and measurement direction.

The general approach is to place the ball rod in four vertical positions, eight horizontal positions, and eight positions at a 45° angle, relative to the CMA and measure five points for each sphere to calculate the average point values. This results in 20 distance values corresponding to 20 measurement positions.

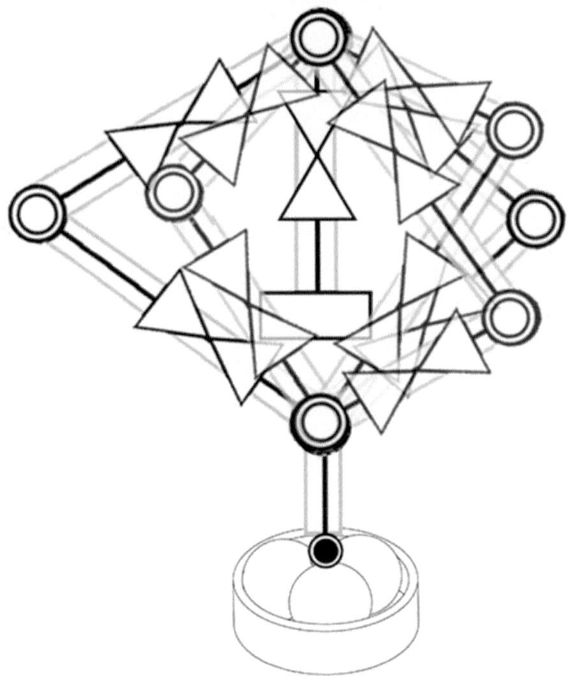

Fig. 2.7 Positioning sequence for determining a single-point position

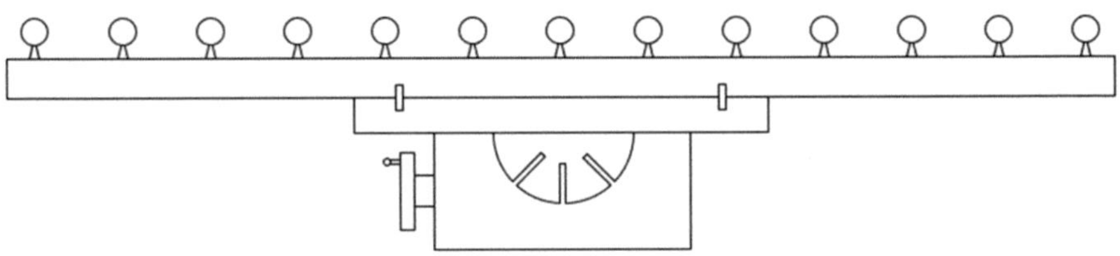

Fig. 2.8 Ball bar

Since the entire calibration process involves three different measurement procedures, specific sources of uncertainty must be considered separately for each test. For the effective diameter test, these sources are the calibrated diameter of the calibration sphere, temperature changes during the measurement, and uncertainty due to insufficient rigidity of the calibration sphere mount [14]. In a single-point measurement test, the main sources of uncertainty are the stiffness of the triangular socket and its geometric accuracy [15].

The geometric accuracy of the socket is controlled by a coordinate measuring machine, and the uncertainty in the stiffness is assumed to be zero. For the volumetric measurement test, the sources of uncertainty are the calibrated size of the ball

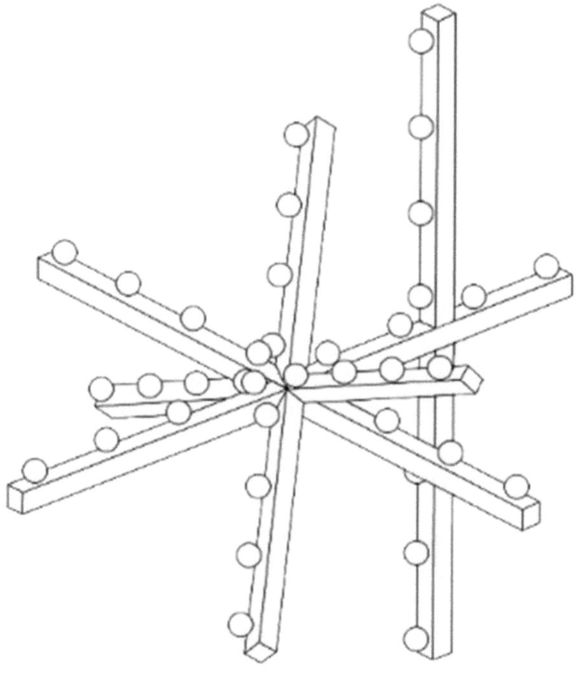

Fig. 2.9 Positioning of the ball bar

rod (according to the calibration certificate), the dimensional drift of the ball rod, temperature changes, and mounting uncertainty [16].

For each of these tests, the combined standard uncertainty is calculated as the positive square root of the sum of the squares of these individual component uncertainties.

$$u_c(y) = \sqrt{\sum_{i-1}^{n} u_i^2},$$

where $u_c(y)$—combined standard uncertainty; u_i—individual uncertainty components.

The expanded uncertainty for each case is obtained by multiplying the combined standard uncertainty $u_c(y)$ by the coverage factor k. For a confidence level of 95%, the coverage factor is usually approximately 2 [17].

The use of the proposed approach will help to improve the accuracy of measurements of geometric parameters of power equipment parts, thus ensuring more reliable and efficient production processes.

2.4 Method for Correcting Dynamic Deviations Based on the Actual Position of Segments

Coordinate measuring arms are becoming increasingly popular in various manufacturing environments due to their simple design, convenient portability, and flexibility in operation, as described in. In general, the factors that can affect CMA accuracy can be divided into static and dynamic.

Static factors focus on the deviation between the actual parameters, which can affect the measurement accuracy with various predictable errors, and the theoretical parameters. To improve the measurement accuracy and reduce the influence of static factors on the CMA, methods such as structure parameter modelling, calibration and error compensation techniques are used.

However, the accuracy of CMA can be affected by both static and dynamic factors. Static factors refer to the predictable errors arising from deviations between actual and theoretical parameters and can be reduced by error compensation techniques such as structure parameter modelling [18], calibration and error compensation. On the other hand, there are dynamic factors related to errors that arise due to elastic deformation, inertia and weight of connections, space between connections, and encoder hopping.

To compensate for dynamic factors, it is necessary to investigate the nonlinear relationships between various errors that affect accuracy. However, individual errors are difficult to separate, so it is proposed to consider all factors together. In the case of CMAs, the configuration is determined by the motion parameters, which include the length of the rods and the values of the encoders in different joints. Usually, the dynamic factors of the CMA can be related to the configuration, which is determined by the values of the encoders of all joints [19].

Provided that encoder readings are available, error compensation in various configurations of the CMA can be implemented by modelling the system. Based on the conclusions of, we analysed the dependence between the arm bending angles and the probe coordinates. To improve the effectiveness of residual error correction, a triangular model of the measuring arm was studied in, taking into account the characteristics of the CMA. A polynomial model of residual error correction was proposed, taking into account the complex error compensation based on the triangular model.

Let's consider a method of residual error correction based on the configuration of the position of the coordinate measuring arm, which can significantly improve the measurement accuracy. The single-point residual kinematic error correction model is used to improve the accuracy of the CMA and is based on the configuration of the CMA position and includes a method for correcting single-point residual errors using a polynomial model. Let us consider the method of correcting the residual kinematic error of the CMA based on the configuration of single-point positions. As an example, we use a 6-element CMA (Fig. 2.10), the configuration of which is determined by the angles of all connecting elements of the system (Fig. 2.11), and the position matrix P can be expressed by the formula

$$P = \prod_{j=1}^{n} A(\theta_j), \tag{2.12}$$

where $A(\theta_j)$ is the Denavit-Hartenberg parameter matrix for the *j-th* connecting element, which is determined by the value of the angle θ_j of this connecting element.

Since the terminal connectors of the CMA, which include the fifth and sixth connectors, are short in length and light in weight, the impact on the system from them can be ignored.

The main elements of CMA quality are the structural elements from the second to the fourth connecting element, which are important parts that lead to a certain regularity in the accuracy of CMAs in different position configurations.

As a result, the relationship between the angles of the connecting elements (θ_1, θ_2, θ_3) and the coordinates of the probe exists in the correction of residual errors.

As shown in Fig. 2.11, a set of configurations is established when the position of the CMA probe with the first four connecting elements (θ_1, θ_2, θ_3) is determined. In

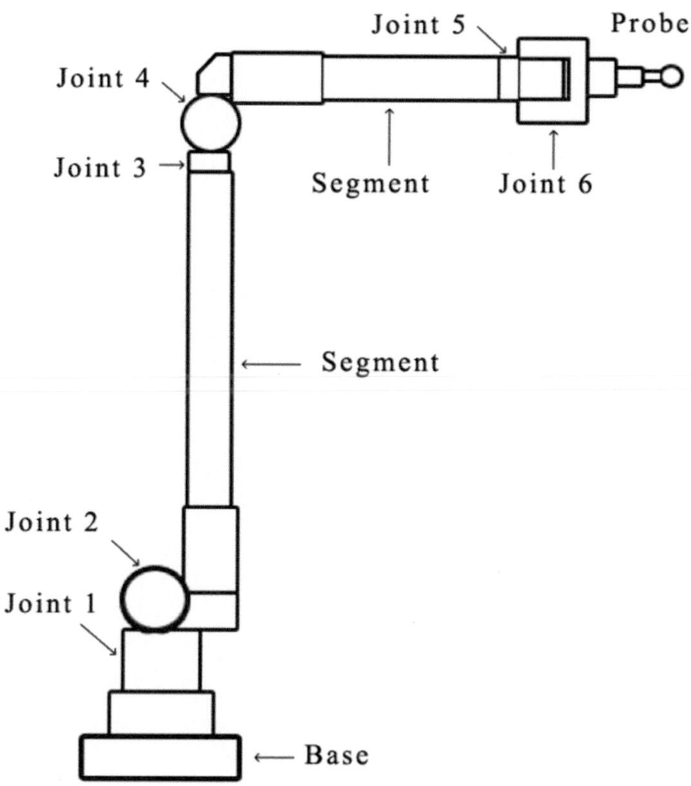

Fig. 2.10 6-element CMA

Fig. 2.11 Configuration of a 6-element CMA

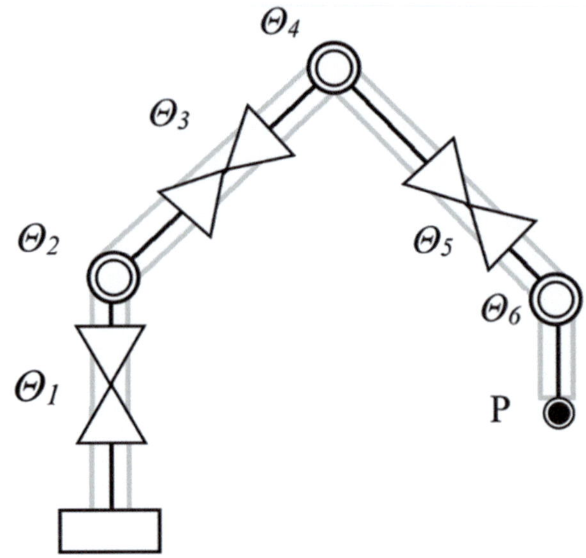

theory, the set of configurations can be located on the boundary of a circle called the configuration circle (Fig. 2.12).

The coordinate of the O_5 connection point $(x_{O_5}, y_{O_5}, z_{O_5})$ can be calculated from the inverse solution of the Denavit-Hartenberg parameter matrix, as described in [20].

Fig. 2.12 Diagram of the layout of the CMA elements

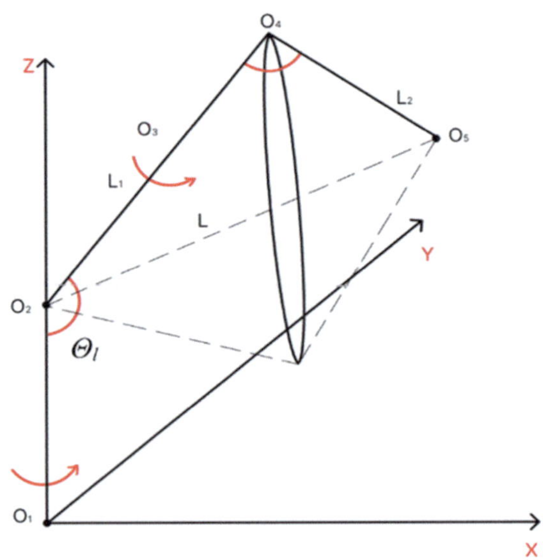

Given the position of O_5 ($x_{O_5}, y_{O_5}, z_{O_5}$), we can calculate the coordinate of O_2 ($x_{O_2}, y_{O_2}, z_{O_2}$), the lengths L_1, L_2, the distance between O_2 and O_5, which is defined as L, by the formula:

$$L = \sqrt{\left(x_{O_2} - x_{O_5}\right)^2 + \left(y_{O_2} - y_{O_5}\right)^2 + \left(z_{O_2} - z_{O_5}\right)^2}. \tag{2.13}$$

The value of the angle θ_4 for the fourth connection can be calculated by the formula:

$$\cos\theta_4 = \frac{L_1^2 + L_2^2 - L^2}{2L_1 L_2}. \tag{2.14}$$

According to formulas (2.13) and (2.14), the position of the triangle Δ O_2, O_4, O_5 in the O_1 coordinate system (x, y, z) can be determined using the angle σ_3 of the third connecting element and the angle θ_l between the Z-axis and the linear segment O_2 O_5.

As shown in Fig. 2.13, the angle θ_l can be calculated using the formulas:

$$\theta_l = \pi - \arccos\left(\frac{|ab|}{|a||b|}\right); \tag{2.15}$$

$$a = O_i + O_j + z_{O_2} k; \tag{2.16}$$

$$b = \left(x_{O_5} - x_{O_2}\right)i + \left(y_{O_5} - y_{O_2}\right)j + \left(z_{O_5} - z_{O_2}\right)k, \tag{2.17}$$

where a is the direction vector of the z-axis; b is the direction vector of the segment O_2 O_5. As a rule, the z-axis is perpendicular to the horizontal plane [21–23].

Let us consider the dynamic deformation of the CMA as parameters Δz, Δs and Δh. As shown in Fig. 2.14, Δz is the Z-coordinate O_5 strain; Δs is the tangential O_5

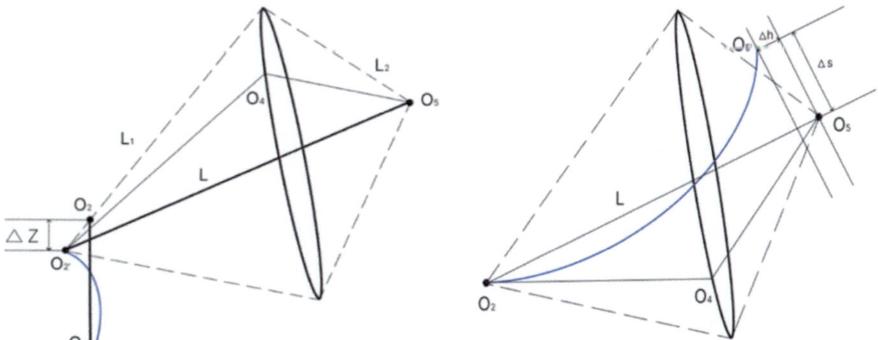

Fig. 2.13 Dynamic deformation of the CMA: **a** front view, **b** deformation view from the top

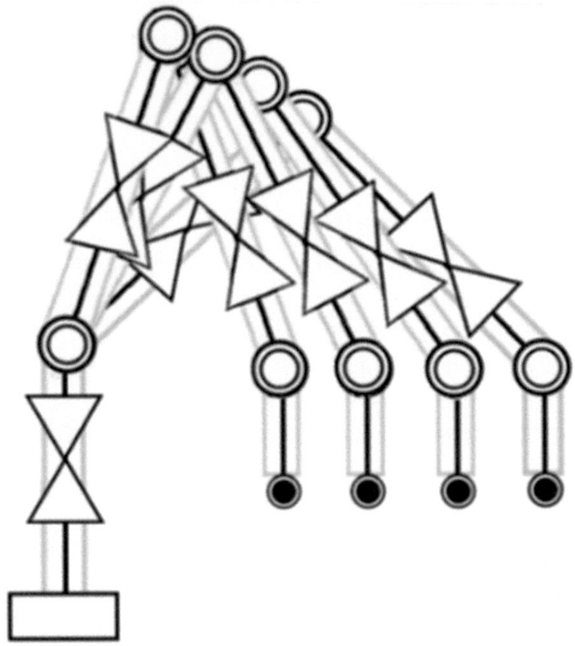

Fig. 2.14 Definition of a single point position

strain in the XY-plane; Δh is the radial O_5 strain in the XY-plane. (Δz, Δs, Δh) can be called residual errors in the cylindrical coordinate system of the CMA, which are the errors between theoretical and practical values.

Therefore, the position configuration depends on θ_3, θ_4 and θ_l and the mapping relation between (θ_3, θ_4, θ_l) and (Δz, Δs, Δh) can be expressed by:

$$f(\theta_3, \theta_4, \theta_l) \rightarrow (\Delta \mathrm{z}, \Delta \mathrm{s}, \Delta \mathrm{h}). \tag{2.18}$$

The residual correction is calculated by approximating Eq. (2.18) using a polynomial [24]:

$$\begin{aligned}
\Delta \mathrm{z} &= k_{z_0} + \sum_{i=0}^{2}\left(k_{z_{(1+3i)}}\theta_3^i\theta_4^{2-i} + k_{z_{(2+3i)}}\theta_4^i\theta_l^{2-i} + k_{z_{(3+3i)}}\theta_l^i\theta_3^{2-i}\right);\\
\Delta \mathrm{s} &= k_{s_0} + \sum_{i=0}^{2}\left(k_{s_{(1+3i)}}\theta_3^i\theta_4^{2-i} + k_{s_{(2+3i)}}\theta_4^i\theta_l^{2-i} + k_{s_{(3+3i)}}\theta_l^i\theta_3^{2-i}\right);\\
\Delta \mathrm{h} &= k_{h_0} + \sum_{i=0}^{2}\left(k_{h_{(1+3i)}}\theta_3^i\theta_4^{2-i} + k_{h_{(2+3i)}}\theta_4^i\theta_l^{2-i} + k_{h_{(3+3i)}}\theta_l^i\theta_3^{2-i}\right),
\end{aligned} \tag{2.19}$$

where, k_{z_j}, k_{s_j}, k_{h_j} are the coefficients of the polynomial.

Thus, the formulated analytical relationships between configuration angles, dynamic deviations, and systematic residual errors form the basis of a correction method that takes into account the actual spatial position of the CMA segments and their deformation.

2.5 Construction of a Single-Point Model of Residual Kinematic Error Correction

The developed model for correcting residual kinematic errors allows determining the discrepancies between theoretically calculated and actual coordinates of the CMA probe that arise due to kinematic errors. The use of polynomial approximation makes it possible to calculate the correction factors for the coordinates of the measured points based on the configuration parameters of the CMA, which, in turn, contributes to the CMA accuracy.

The method of building a residual correction model consists of the following steps:

1. Data collection. The horizontal plane of the base is taken as a reference in the measuring space of the CMA, and several positions are selected within 20–80% of the radial direction of the CMA measurement (Fig. 2.14). At these locations, using the method shown in Fig. 2.15, data are collected from a single point. The data θ_3 and θ_4, which are the values of the O_3 and O_4 connections of the CMA, can be obtained from the encoders. The coordinates of the probe and L can be calculated using Eqs. (2.14) and (2.15).
2. The average value $\overline{x}, \overline{y}, \overline{z}$ and the coordinates of one point (x_i, y_i, z_i) are used as initial data for calculating the residual values. Then (Δz_i, Δs_i, Δh_i) can be expressed by formula (2.20):

$$\begin{aligned}
&\Delta z_i = z_1 - \overline{z};\\
&\Delta s_i = \sqrt{x_i^2 + y_i^2}\sin\theta_i;\\
&\Delta h_i = \sqrt{x_i^2 + y_i^2}\cos\theta_i - \sqrt{\overline{x}^2 + \overline{y}^2};\\
&\theta_i = arc\tan\tfrac{\overline{y}}{\overline{x}} - arc\tan\tfrac{y_i}{x_i}.
\end{aligned} \tag{2.20}$$

3. According to (2.19), the residual correction equations can be established together with ($\theta_3, \theta_4, \theta_l$) and ($\Delta z_i, \Delta s_i, \Delta h_i$). And the polynomial coefficients ($k_{z_j}, k_{s_j}, k_{h_j}$) can be calculated from the residual correction equations.
4. The compensation value ($\Delta x_i, \Delta y_i, \Delta z_i$) for the coordinate ($x_i, y_i, y_i$) of the CMA probe was calculated using the residual value correction method. In accordance

with the relationship between the cylindrical coordinate system and the Cartesian coordinates, the compensation value can be calculated by Eq.

$$\begin{cases} \Delta x_i = |\Delta h_i| \cos\theta_i + |\Delta s_i| \sin\theta(\Delta s_i \geq 0) \\ \Delta x_i = |\Delta h_i| \cos\theta_i - |\Delta s_i| \sin\theta(\Delta s_i < 0) \end{cases};$$

$$\begin{cases} \Delta y_i = |\Delta s_i| \cos\theta_i + |\Delta h_i| \sin\theta(\Delta h_i < 0) \\ \Delta y_i = |\Delta s_i| \cos\theta_i - |\Delta h_i| \sin\theta(\Delta h_i \geq 0) \end{cases}; \tag{2.21}$$

$$\begin{aligned} &x = x_i - \Delta x_i, \quad x = x_i + \Delta x_i; \\ &y = y_i + \Delta y_i(x < 0), y = y_i - \Delta y_i(x_i \geq 0); \\ &z = z_i - \Delta z_i, \quad z = z_i + \Delta z_i; \end{aligned} \tag{2.22}$$

where

$$\theta = arc\tan\frac{y_i}{x_i} + arc\sin\frac{\Delta s_i}{\sqrt{x_i^2 + y_1^2}}.$$

Then the mean square error (MSE) can be calculated as (2.23) and (2.24)

$$\delta_i = \sqrt{(x_i - x_a)^2 + (y_i - y_a)^2 + (z_i - z_a)^2}; \tag{2.23}$$

$$MSE = \frac{1}{n}\sum_{i=1}^{n} \delta_i^2, \tag{2.24}$$

where n is the number of single-point measurements; (x_i, y_i, z_i) are coordinates for each measurement; (x_a, y_a, z_a) is the average value of measurements over n times [25].

The algorithm of calculation by the residual values correction method is shown in Fig. 2.16.

The application of this method allows to significantly reduce the residual kinematic errors of the CMA. The residual kinematic errors were corrected using computer modelling. The structural parameters of the 6-element CMA are given in Table 2.1.

10 single-point measurements were made using the method shown in Fig. 2.6. The results are shown in Table 2.2.

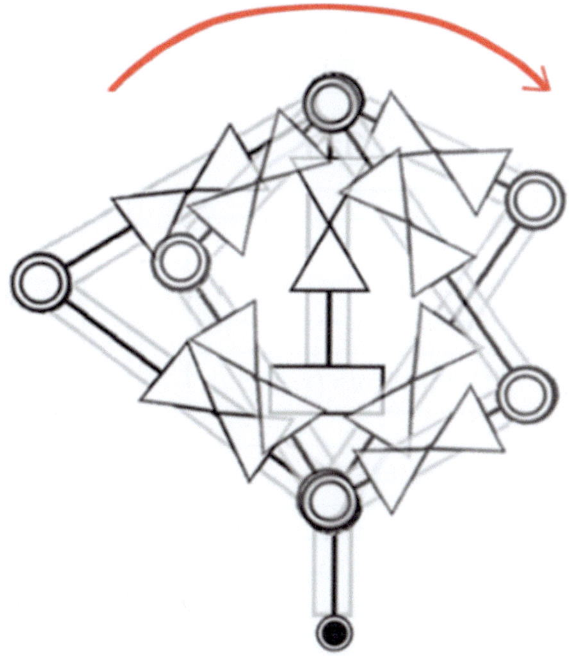

Fig. 2.15 Positioning sequence for determining a single-point position

The data obtained after applying the method of single-point correction of residual kinematic errors are given in Table 2.3.

Comparative characteristics of the results of computer modelling of the measurements before and after applying the method of single-point correction of residual kinematic errors are shown in Figs. 2.17 and 2.18.

To evaluate the effectiveness of the method of single-point correction of residual kinematic errors, the root mean square error (RMSE) was calculated based on the experimental data before and after the correction method was applied. The results of the calculations showed that the RMSE before the application of the method was approximately 0.575 mm, and after its application it decreased to 0.14 mm. Thus, the application of the proposed method of single-point correction of residual kinematic errors of errors allowed to reduce the kinematic error by about 4.09 times.

This method significantly improves the accuracy of measurements of the geometric parameters of power equipment parts, ensuring more reliable and efficient production processes.

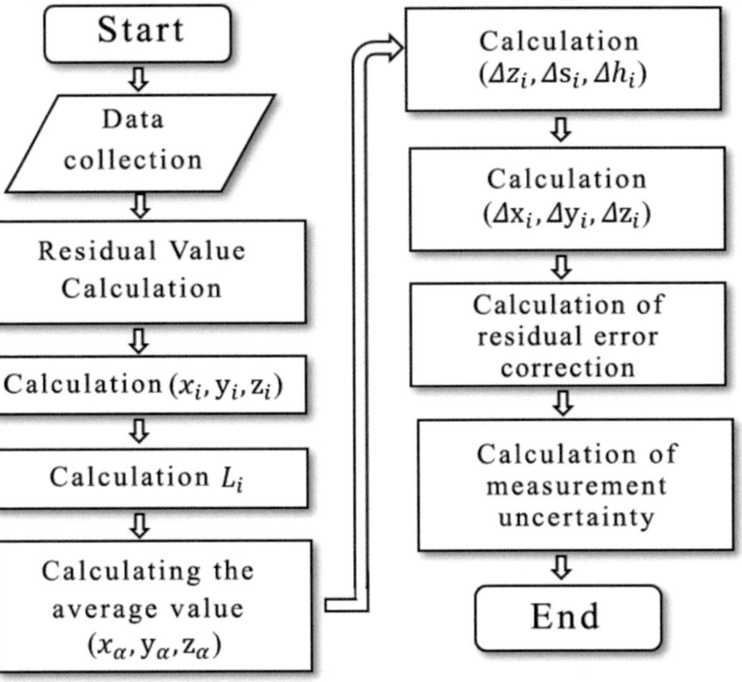

Fig. 2.16 Calculation algorithm using the one-point correction method for residual kinematic errors

Table 2.1 Denavit-Hartenberg parameters of the 6-element CMA

	Hinge 1	Hinge 2	Hinge 3	Hinge 4	Hinge 5	Hinge 6
l_i (mm)	0	−90	90	0	0	155.479
d_i (mm)	0	0	−650	0	−435	65.931
a (i)	−1.571	1.571	−1.571	1.571	1.571	0
θ_i (rad)	0	0.001	0.002	0.003	0.005	0.04

Table 2.2 Results of single-point measurements before applying the method

No.	1	2	3	4	5	6	7	8	9	10
x (mm)	0.231	0.182	0.182	0.083	0.053	−0.128	−0.153	−0.311	−0.086	−0.093
y (mm)	0.611	0.554	0.421	0.361	0.116	−0.421	−0.486	−0.429	−0.386	−0.283
z (mm)	−0.263	−0.278	−0.335	−0.359	−0.299	−0.063	0.196	0.396	0.479	0.524

Table 2.3 Results of single-point measurements after applying the method

No.	1	2	3	4	5	6	7	8	9	10
x (mm)	0.094	0.018	0.022	−0.052	0.001	0.022	0.001	−0.178	0.058	0.051
y (mm)	0.098	0.053	−0.026	−0.023	−0.123	−0.099	−0.065	0.026	0.108	0.183
z (mm)	0.042	0.040	−0.012	−0.013	0.042	0.043	−0.120	−0.052	0.009	0.178

2.6 Conclusions

The second chapter of the thesis presents an integrated approach to improving the accuracy of measurements of geometric parameters of power equipment parts. The kinematic model of the information and measuring system based on the coordinate measuring arm was improved, which allowed to take into account the real deviations of parameters, in particular, the lengths of links, angles between them and zero shifts of encoders. A combined calibration method combining an effective diameter test, a single-point measurement test, and a volumetric measurement test within a single procedure was proposed, which provides a systematic account of kinematic errors in and contributes to the repeatability, accuracy, and reproducibility. A model of single-point correction of residual kinematic errors using polynomial approximation has been developed, which allows reducing residual kinematic errors by four times [26]. This set of methods demonstrates a significant improvement in the accuracy and reproducibility of measurements, which is of great practical importance for the use of high-precision information and measurement systems in the energy and mechanical engineering industries.

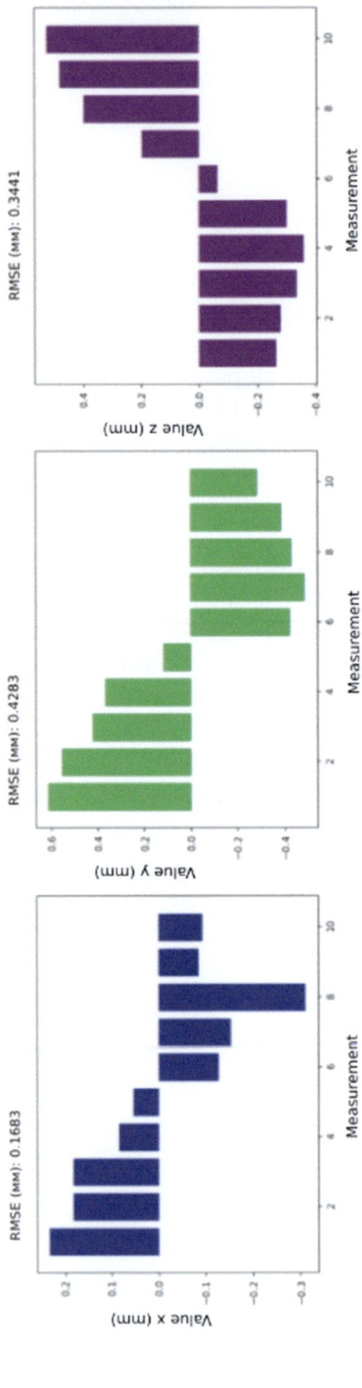

Fig. 2.17 Before applying the single-point correction method

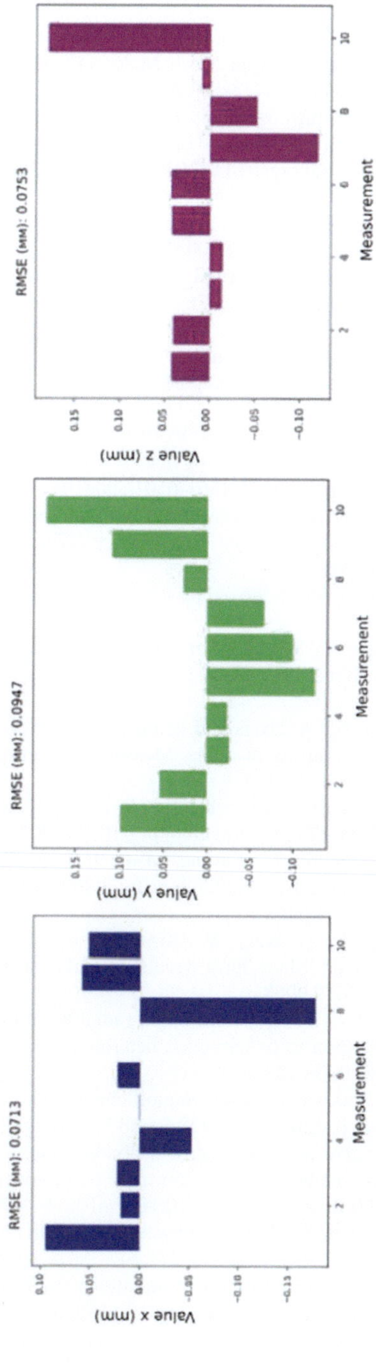

Fig. 2.18 After applying the single point correction method

References

1. Xiong, H., Xu, Y., Zong, Z., Gao, Q.: The multi-joint measuring arm's error compensation method based on the pose configurations. Appl. Mech. Mater. **117–119**, 751–755 (2012). https://doi.org/10.4028/www.scientific.net/amm.117-119.751
2. Lopez-Franco, C., Hernández-Barragán, J., Alanis, A.Y., Arana-Daniel, N., López-Franco, M.: Inverse kinematics of mobile manipulators based on differential evolution. Int. J. Adv. Robot. Syst. **15**(1), 27–38 (2018). https://doi.org/10.1177/1729881417752738
3. Bing, H., Linxian, C., Chusheng, L., Yonggang, H.: Complex differential evolution algorithm for inverse kinematics of spatial 6R robot manipulators. J. Mech. Eng. **50**(15), 45–52 (2014) http://www.cjmenet.com.cn/CN/Y2014/V50/I15/45
4. Gao, G., Zhang, H., Wu, X., Guo, Y.: Structural parameter identification of articulated arm coordinate measuring machines. Math. Probl. Eng. **3**, 1–10 (2016). https://doi.org/10.1155/2016/4063046
5. Babak, S., Babak, V., Zaporozhets, A., Sverdlova, A.: Method of statistical spline functions for solving problems of data approximation and prediction of objects state. SSRN Electron. J. (2019). https://doi.org/10.2139/ssrn.3734104
6. Asmai, E., Hennebelle, F., Coorevits, T., Rénald, V., Fontaine, J.-F.: Rapid and robust on-site evaluation of articulated arm coordinate measuring machine performance. Meas. Sci. Technol. **29**(11), 1–14 (2018). https://doi.org/10.1088/1361-6501/aade10
7. Zaporozhets, A., Kataiev, D.: Method of compensating for instrumental uncertainty in measurements using a coordinate measuring arm. Syst. Res. Energy **1**(76), 45–53 (2024). https://doi.org/10.15407/srenergy2024.01.045
8. ISO/IEC Guide 98-1:2024(en) Guide to the expression of uncertainty in measurement—Part 1: Introduction
9. Gao, G., Sun, G., Na, J., Guo, Y., Wu, X.: Structural parameter identification for 6 DOF industrial robots. Mech. Syst. Signal Process. **113**, 145–155 (2018). https://doi.org/10.1016/j.ymssp.2017.08.011
10. Luo, Z., Liu, H., Li, D., Tian, K.: Analysis and compensation of equivalent diameter error of articulated arm coordinate measuring machine. Measurem. Control. **51**(5), 16–26 (2018). https://doi.org/10.1177/0020294018755324
11. Babak, S., Babak, V., Myslovych, A., Zaporozhets, A., Zvaritch, V.: Simulation and software for diagnostic systems, pp. 71–90 (2020). https://doi.org/10.1007/978-3-030-44443-3_3
12. Daniel, G.-M., Joaquín, B., Eduardo, C., Susana, M.-P.: Influence of human factor in the AACMM performance: a new evaluation methodology. Int. J. Precis. Eng. Manuf. **15**(7) (2014). https://doi.org/10.1007/s12541-014-0468-9
13. Gao, G., Zhang, H., San, H., Wu, X., Wang, W.: Kinematic calibration for industrial robots using articulated arm coordinate machines. Int. J. Model. Identif. Control. **31**(1), 16–26 (2019). https://doi.org/10.1504/IJMIC.2019.096816
14. Park, B.-J., Pedrycz, W., Oh, S.-K.: Polynomial-based radial basis function neural networks (P-RBF NNs) and their application to pattern classification. Appl. Intell. **32**, 27–46 (2010). https://doi.org/10.1007/s10489-008-0133-z
15. Eremenko, V.S., Kuts, Y.V., Mokiychuk, V.M., Samoilichenko, O.V.: Statistical Analysis of Measurement Data: A Tutorial, p. 320. National Aviation University (2013)
16. Sagara, S., Taira, Y.: Cooperative manipulation of a floating object by some space robots: application of a tracking control method using the transpose of the generalized Jacobian matrix. Artif. Life Robot. **12**(1–2) (2008). https://doi.org/10.1007/s10015-007-0455-7
17. Bosch, J.A.: Coordinate Measuring Machines and Systems, 2nd edn (2016). https://doi.org/10.1201/b11022
18. Wang, W., et al.: Analysis and compensation of installation errors for circular grating angle sensors. Adv. Sci. Lett. **4**(6–7) (2011). https://doi.org/10.1166/asl.2011.1633
19. Elatta, A.Y., Gen, L.P., Zhi, F.L., Daoyuan, Y., Fei, L.: An overview of robot calibration. Inform. Technol. J. **3**, 74–78 (2004). https://doi.org/10.3923/itj.2004.74.78

20. Santolaria, J., Brau, A., Velázquez, J., Aguilar, J.J.: A self-centering active probing technique for kinematic parameter identification and verification of articulated arm coordinate measuring machines. Meas. Sci. Technol. (2010). https://doi.org/10.1088/0957-0233/21/5/055101
21. Piratelli-Filho, A., Lesnau, G.R.: Virtual spheres gauge for coordinate measuring arms performance test. Measurement 236–244 (2010). https://doi.org/10.1016/j.measurement.2009.10.002
22. Piratelli-Filho, A., Henrique, F., Fernandes, T., Valdés, R.: Application of virtual spheres plate for AACMMs evaluation. Precis. Eng. **36**, 349–355 (2012). https://doi.org/10.1016/j.precisioneng.2011.10.004
23. Kovac, I., Frank, A.: Testing and calibration of coordinate measuring arms. Precis. Eng. **25**, 90–99 (2001). https://doi.org/10.1016/S0141-6359(00)00057-X
24. Xi, F., Liu, Y., Feng, H.-Y.: Error compensation for three-dimensional line laser scanning data. Int. J. Adv. Manuf. Technol. **18**, 211–216 (2001). https://doi.org/10.1007/s001700170076
25. Forest, J., Salvi, J.: A review of laser scanning three-dimensional digitisers. In: Proceedings of IEEE Conference on Intelligent Robots and Systems. Lausanne, Switzerland, pp. 73–78 (2002)
26. Schwenke, H., Knapp, W., Haitjema, H., Weckenmann, A., Schmitt, R., Delbressine, F.: Geometric error measurement and compensation of machines—an update. CIRP **57**, 660–675 (2008). https://doi.org/10.1016/j.cirp.2008.09.008

Chapter 3
Software and Information System for Coordinate Measuring Arms with Neural Network Error Compensation

3.1 Introduction

The method of single-point correction of residual kinematic errors proposed in the previous section in combination with the combined method of CMA calibration demonstrates a significant reduction in kinematic errors. At the same time, solving the problem of improving the accuracy of measurements of geometric parameters of power equipment parts also requires effective compensation of nonkinematic errors that cannot be compensated by classical methods.

To achieve this goal, it is proposed to create a specialised software and information system capable not only of automating the processes of collecting, processing and analysing measurement data, but also of implementing adaptive compensation of nonkinematic errors caused by the influence of external destabilising factors. Implementation of such compensation requires the introduction of modern intelligent algorithms, in particular machine learning methods, which are capable of detecting complex patterns in the structure of residual errors and correcting them.

The third chapter of the book is devoted to the development of the structure and functional content of such a complex. Within this chapter, the authors propose a structural scheme of a software and information system with an integrated neural network, defines the criteria for choosing the architecture of a neural network, and substantiates the feasibility of using specific software tools for building, training and integrating NNs into a software and information system.

D. Kataiev et al., *Neural Calibration of Coordinate Measuring Arms*, Studies in Systems, Decision and Control 695, https://doi.org/10.1007/978-3-032-32918-9_3

3.2 Development of a Structural Diagram of a TDF with an Integrated Neural Network

The information and measurement system consists of several interconnected components (Fig. 3.1), which together provide high-precision control and adaptive data processing [1–4].

The software and information system (SIS) is the central control component that determines the order of reading signals from encoders, provides data preparation, regulates the launch of processing algorithms, controls the neural network, interacts with the database and provides access to the user interface. The software and information system (Fig. 3.2) consists of the following main components: system core, calibration module, measurement control module, data storage module, primary data collection module, spatial coordinates calculation module, internal resources management module, graphical interface, diagnostics module, access control module.

User interface (Fig. 3.3): main screen, system status panel, settings panel, quick access panel, notification panel, menu panel, measurement panel, diagnostics panel, NN configuration panel.

The user interface allows the operator to initiate a measurement sequence, select the desired operating mode (Figs. 3.4 and 3.5), and receive results in the form of coordinates, deviation statistics, and messages about possible technical warnings.

The measurement process is as follows. Touch sensor detects the moment of contact with the object's surface and, taking into account the angles received from the encoders, sends the corresponding signal to the data processing unit, where the coordinates of (X, Y, Z) probe are calculated.

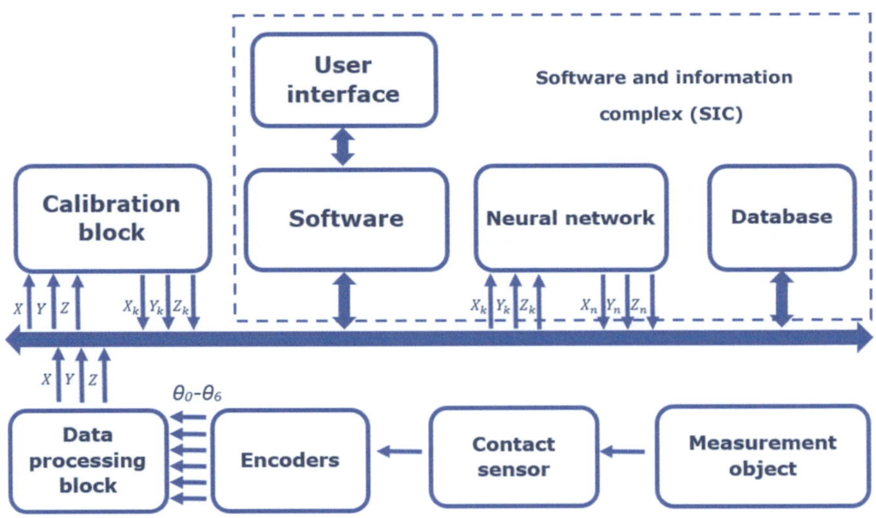

Fig. 3.1 Structural scheme of IMS

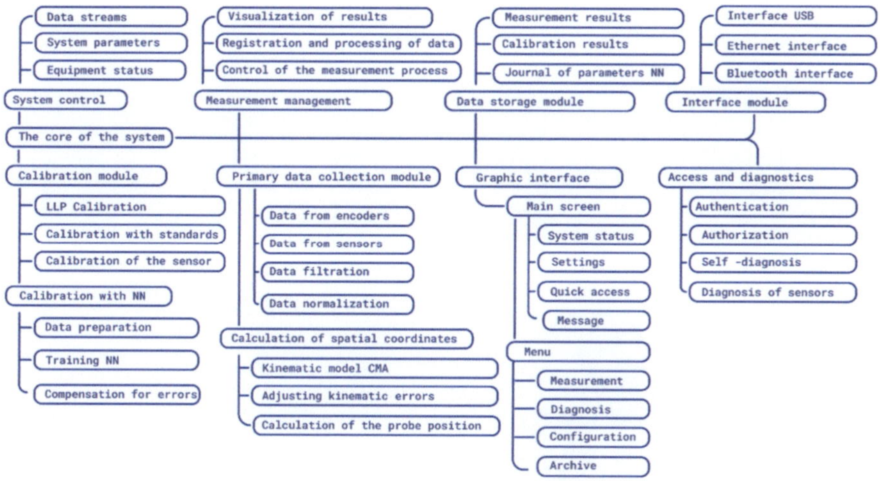

Fig. 3.2 Structural scheme of the software and information system

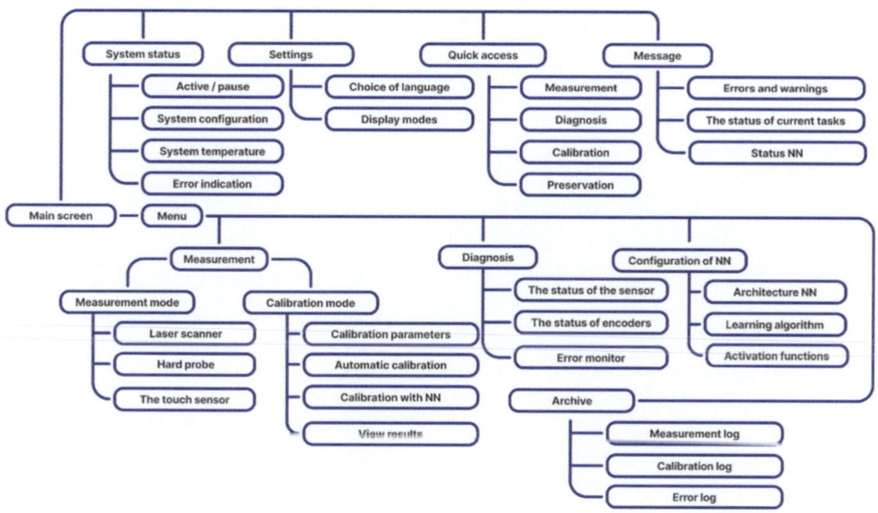

Fig. 3.3 Interface map of the SIS

To ensure reliable synchronisation of signals coming from encoders, it is proposed to add a pulse follower frequency multiplier based on phase-locked automatic frequency adjustment (PFA) to the hardware of the CMA (Fig. 3.6). The peculiarity of such a multiplier is the presence of an additional analogue control loop, which reduces phase noise and allows expanding the range of processed frequencies [5].

Figure 3.6 shows the following notations: 1—period-to-voltage converter, 2—pulse modulator, 3—memory element, 4—controlled oscillator, 5—frequency

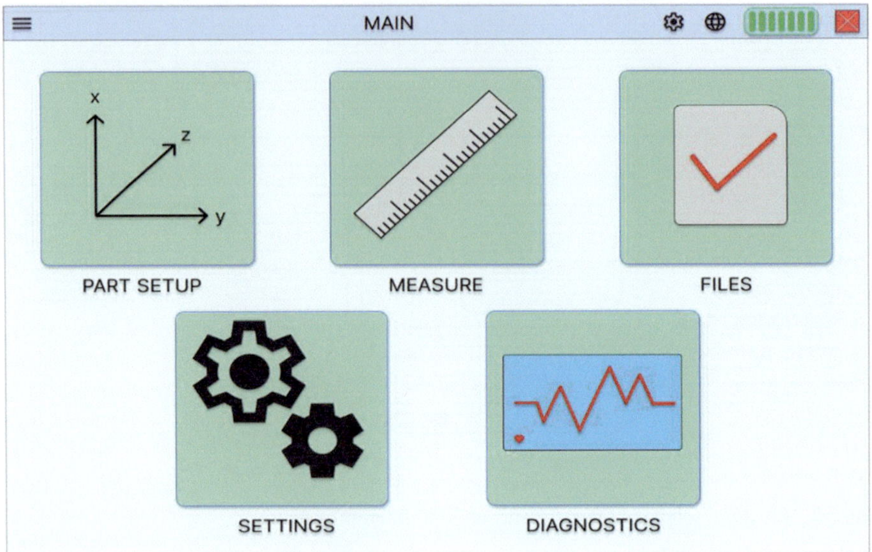

Fig. 3.4 User interface (home page)

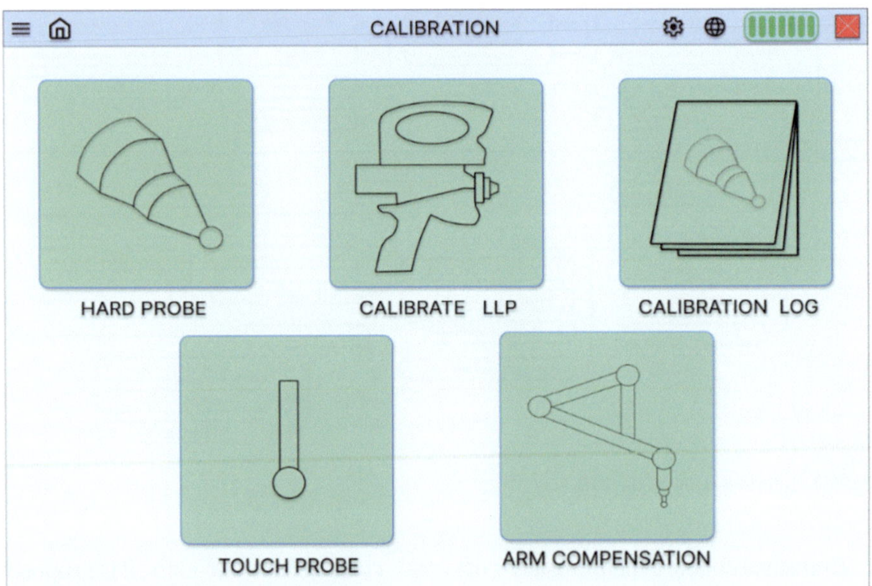

Fig. 3.5 User interface (calibration page)

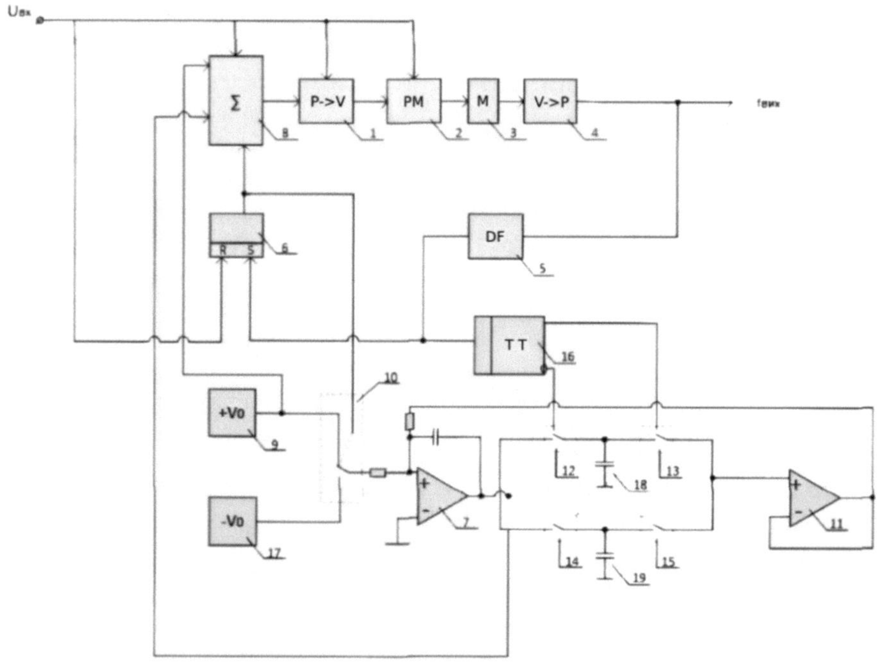

Fig. 3.6 Pulse train frequency multiplier

divider, 6—phase comparator trigger, 7—integrator, 8—adder, 9—positive source of constant reference voltage, 10—automatic switch, 11—voltage repeater, 12, 13, 14, 15—analogue switches, 16—T-trigger, 17—negative source of constant reference voltage.

The period-to-voltage converter (1) is the first element that receives the input pulse sequence and converts its period into a linearly varying voltage. The input pulse frequency is then displayed as an analytically controlled voltage level, which is then used for further modulation and control. The pulse modulator (2) is also synchronised to the input pulses. It creates short pulse "slices" of the signal, which record the instantaneous voltage value from the converter (1). This allows timely transfer of information about the current period to the memory element. The memory element (3) receives the corrected voltage from the pulse modulator (2) and stores it for the entire period until the next synchronisation cycle arrives. This reduces the instantaneous fluctuations of the signal and provides a "steady state" voltage at the input of the controlled oscillator. The controlled oscillator (4) generates a frequency output signal influenced by the stored level from the memory element (3). By changing the control voltage, the output frequency can be changed. This performs the main task of multiplying or adjusting the frequency relative to the input pulse sequence. The frequency divider (5) divides the frequency generated by the oscillator (4) by a setpoint, which is closed in a feedback loop. It feeds the comparison pulses back into the phase comparator and thus ensures the correct ratio between the input and output signals. The trigger of the phase comparator (6) receives two pulse signals: one directly from the input sequence and the other from the output of the frequency divider (5). By comparing them in phase, the trigger (6) generates a phase error signal that affects the slope of the converter (1) and, ultimately, regulates the correctness of the synchronisation. The integrator (7) accumulates (integrates) the difference signal received from the phase comparator trigger (6) via other control links. In this way, a slow (astatic) phase correction is formed, which eliminates long-term deviations and ensures stability and noise reduction. The adder (8) adds two signals: an input from the integrator (7) and a constant reference voltage from the source (9). This allows you to set the basic control level of the converter (1) and increase or decrease the influence of the integral component. The constant reference voltage source (9) provides the positive reference signal used in the adder (8) and in the automatic switching logic (10). It is this source that determines the initial or upper limit by which the control level will be shifted. The automatic switch (10) selects between the two reference voltage sources (positive (9) and negative (17)) under the influence of the control signal from the phase comparator trigger (6). This makes it possible to accelerate or decelerate the charge/discharge of the integrator (7) depending on the sign of the phase error. A voltage follower (11) with a low output impedance ensures stable signal transmission between the integrator (7) and the analogue switching units (12, 13, 14, 15). It acts as a buffer to reduce the influence of the load on the voltage level. The analogue switches (12, 13, 14, 15) are controlled by a T-trigger (16) and alternately connect the output of the integrator (7) to the repeater (11) and, respectively, to capacitors that are grounded. This alternate switching allows changing the charging and discharging pattern of the integrator stage, which further smoothes out short-term impulse noise and contributes to better phase stabilisation. The T-trigger (16) is connected to the frequency divider (5) with its counting input and generates variable control states for the analogue switches (12, 13, 14, 15). Depending on the clock pulses, one output of the T-trigger opens a certain combination of keys, and the other closes them, thus alternately organising the charging or holding mode of the signal in the integrator. A negative constant reference voltage source (17) provides the lower level of the reference signal used by the automatic switch (10). This enables the system to control the phase correction signs by switching the integrator (7) between a positive and negative reference potential.

The principle of operation of the pulse train frequency multiplier is based on comparing the input and the return (divided) frequency and automatically adjusting the controlled oscillator to a multiple of the input signal. If we denote the input frequency by f_i, and the output (multiplied) frequency by f_o, the circuit forms

$$f_o = N \cdot f_i, \tag{3.1}$$

where N is determined by the parameters of the divider and generator in the PFA circuit. The coefficient N can be adjusted by the slope control unit of the period-to-voltage converter.

To explain the principle of integration with other components of the IMS, it is advisable to highlight such functions of the multiplier as phase and frequency stabilisation (smoothing of pulse interference due to the integrator mode), and generation of an output signal with a unified or higher frequency compared to the original encoder signal [6].

It is advisable to use a pulse train frequency multiplier at the stage of preliminary hardware signal processing. This allows you to isolate and amplify the "reference" clock level, reducing the influence of phase fluctuations that occur directly in the joints of the measuring system. A period-to-voltage converter generates a linearly variable voltage during each input interval (corresponding to $1/f_i$), and an integrator with a phase comparator provides additional feedback:

$$U(t) = \int [\alpha \cdot (\phi_i(t) - \phi_o(t))] \, dt, \tag{3.2}$$

where $\phi_i(t)$ and $\phi_o(t)$ are the phases of the input and output signals, and α is the gain.

This integral component ensures the astatic property of the device and works out a zero average phase error, which reduces jitter and increases system performance in the event of sudden changes in the input frequency [7].

In the implementation of the IMS, the integrated pulse train frequency multiplier works as follows. The raw pulse signals are fed from the encoders to the pulse selection and shaping unit. The input sequence for the multiplier is then generated. Thanks to the PFA, the output frequency of f_o becomes more stable: local fluctuations or pulse disturbances from sensors are partially smoothed out in the analogue circuit. Thus, the pulse follower frequency multiplier in the IMS plays the role of a hardware "filter" of phase instability and at the same time a frequency synthesiser, which makes further digital data processing more reliable and accurate.

In the context of coordinate measuring systems, the fact that encoder data is periodic in nature—angles are defined in the range [0°, 360°]—plays an important role, leading to edge effects. For example, the values of 1° and 359° can be interpreted as very distant in a direct calculation, although they are actually close.

To take this feature into account, the Mises distribution is used, a probabilistic model that describes cyclic variables and allows you to correctly determine the statistical characteristics of angular data [8].

$$p_M(\theta \mid \mu, k) = \frac{1}{2\pi I_0(k)} e^{k \cos(\theta - \mu)}, \quad \mid \mu \mid < \infty, \quad k > 0, \tag{3.3}$$

where $I_0(k)$ is a modified Bessel function of the first kind of zero order:

$$I_0(k) = \frac{1}{2\pi} \int_0^{2\pi} e^{k \cos(\theta - \mu)} d\theta, \tag{3.4}$$

where μ is the circular average direction of the random angle; *k is* the concentration parameter of the random angle in the vicinity of μ.

The Kalman filter starts from the stage of predicting the system state. Thanks to preliminary smoothing and correct statistical interpretation of cyclic variables, the Kalman filter works with more reliable and consistent data, which significantly increases the efficiency of error compensation. The input data are the corrected angles $\tilde{\Theta}$, the previous state of the system $\hat{x}_{k-1|k-1}$, and the system dynamics model.

The predicted state is calculated using the formula:

$$\hat{x}_{k|k-1} = F_k \hat{x}_{k-1|k-1} + B_k u_k, \tag{3.5}$$

where $\hat{x}_{k|k-1}$—predicted state of the system, F_k—state transition matrix, B_k—control signal influence matrix, u_k—control influence vector.

At the same time, the error matrix is predicted:

$$P_{k|k-1} = F_k P_{k-1|k-1} F_k^T + Q, \tag{3.6}$$

where $P_{k\,|\,k-1}$ is the predicted error matrix, Q is the process noise covariance matrix.

Based on the new angle measurements from the encoders, the system performs a state update. The measured angles z_k are compared with the predicted values. The innovation (the difference between the measured and predicted data) is calculated as:

$$y_k = z_k - H_k \hat{x}_{k|k-1}, \tag{3.7}$$

where H_k is the observation matrix.

The Kalman coefficient, which determines the weight of new data in the state correction, is calculated as follows:

$$K_k = P_{k|k-1} H_k^T \left(H_k P_{k|k-1} H_k^T + R\right)^{-1}, \tag{3.8}$$

where R is the measurement noise covariance matrix.

The updated state of the system is determined by the formula:

$$\hat{x}_{k|k} = \hat{x}_{k|k-1} + K_k y_k. \tag{3.9}$$

The error matrix is also updated:

$$P_{k|k} = (I - K_k H_k) P_{k|k-1}. \tag{3.10}$$

The output of the Kalman filter is the updated values of the system state, which take into account dynamic effects and are corrected by new data, and the corrected angles of the encoders, $\tilde{\Theta}_k$, which are the result of integrating the dynamics model and measurements [9–12].

This data can be used to calculate the position of the system's endpoint using the direct kinematic model:

$$P = f\left(\tilde{\Theta}_k\right), \tag{3.11}$$

where P is the coordinates of a point in space.

Thus, the proposed method allows taking into account encoder errors and dynamic influences, providing pre-processing and transformation of the received data.

The data processing unit receives the output from the encoders, which contains the measured angle values for each connection in the system. Correction factors received from the calibration unit are applied to the output data. The calibration unit is responsible for performing calibration procedures, compensating for kinematic errors arising from design deviations, such as mismatches in actual link lengths, angular deviations and offsets of encoder readings from nominal values, and generating correction factors that allow the output data to be adjusted to account for kinematic errors.

After data correction, the processing unit uses a forward kinematic model to convert the corrected angular values into the x, y, z coordinates of the probe. This is done by applying a series of transformation matrices, each of which reflects the geometric parameters of a separate segment of the system, such as link length, rotation angle and offset. The result of this process is an accurate calculation of the spatial position of the endpoint (probe) of the system.

However, even after careful compensation of kinematic errors using traditional methods, non-kinematic errors remain, arising from complex nonlinear relationships between system parameters and operating conditions that are difficult to accurately approximate using classical models. Integrating a neural network into the calibration process allows you to adaptively analyse a large amount of data, identify hidden patterns and correct residual deviations, taking into account the influence of numerous factors. Thanks to the ability of the NN to learn from real or synthetic data, the system can automatically update its parameters to minimise residual error, which significantly improves measurement accuracy and reproducibility. Thus, the integration of NNs provides effective compensation for non-kinematic errors, making the calibration process more flexible and adaptive to changing operating conditions.

3.3 Criteria for Determining the Architecture of a Neural Network

Neural networks have become widely used in prediction, recognition and compensation of complex errors, especially when traditional analytical models do not provide sufficient accuracy. The main architectures of neural networks include single-layer perceptron, multilayer perceptron, recurrent neural networks, convolutional neural networks, and autoencoders (Table 3.1).

Table 3.1 Basic neural network architectures

Architecture	Main advantages	Main disadvantages	Applied tasks
Single-layer reception desk	Ease of implementation;quick training;small number of parameters;	Inability to approximate nonlinear functions;limited expressiveness;	Linear classification, basic signal filter
Multilayer perceptron (MLP)	Powerful approximation of nonlinear dependencies;versatility;adaptability to different types of input data;	Sensitivity to the choice of architecture;can be retrained;requires careful configuration of hyperparameters;	Error prediction, deformation compensation, regression analysis
Recurrent neural networks (RNN)	Modelling temporal and sequential dependencies;works well with dynamic data;	Hard to learn (damped/exploding gradients);require a large amount of training data;	Temperature drift, vibration oscillations, serial forecasting
Convolutional neural networks (CNN)	Efficient spatial data processing;high accuracy in localisation and feature extraction tasks;	High computational costs;suboptimal for data without spatial structure;	Surface defect detection, thermal image analysis
Auto-encoders	Capable of detecting anomalies;data compression;building a latent representation;	Tendency to reproduce noise;does not always ensure the interpretability of features;	Dimensionality reduction, detection of atypical errors, denoising

The choice of the neural network architecture is an important step in the development of algorithms for processing and compensating errors in measuring systems. In this case, when it is necessary to take into account irregular influences of non-kinematic origin, as well as to keep the amount of computation moderate, it is most appropriate to use a neural network with back propagation of error, which belongs to the types of multilayer perceptrons [13].

The main advantage of the back-propagation neural network is its ability to approximate complex nonlinear relationships between the initial coordinates and the output corrections required to correct the measured position. This network has a number of important features that make it suitable for processing data obtained with the help of the CMA. Firstly, it provides a simple and generally accepted training mechanism based on the back-propagation algorithm, during which data samples are passed through the network, compared with the expected result, and the recorded error is distributed "backwards" across the layers by adjusting the weighting coefficients.

Secondly, such a neural network does not require excessive computing resources with a moderate number of layers and neurons, which makes it possible to integrate it into the software package of an information and measurement system without significant changes to the hardware component. Thirdly, in the case of acquiring new data under operating conditions, the back-propagation neural network can be effectively retrained to adapt to changed joint characteristics, long-term degradation, or inaccuracies [14–16].

The practical implementation of such a network in a SIS is carried out through the formation of a set of training data, where the input parameters can be obtained using the IMS coordinates (X, Y, Z), and the output data are the necessary corrections to the coordinates (ΔX, ΔY, ΔZ). The training algorithm gradually adjusts the weighted network links to minimise the error on these examples. As a result, the network learns to recognise residual, hard-to-predict factors that cannot be eliminated by classical methods [17–22].

Thus, the introduction of a neural network with back-propagation of error allows not only to expand the functionality of the information and measuring system, but also to ensure its adaptability to changing operating conditions and exogenous factors (temperature, signal distortion, uneven wear, random defects). Due to the versatility of the back propagation mechanism, such a network can be easily integrated with the error reduction algorithms already implemented by and complements them, compensating for nonlinear effects that remain outside the scope of classical approaches.

3.4 Justification of Software Tools for Building a Neural Network

In the process of developing a SIS, it is of fundamental importance not only to choose machine learning algorithms, but also software tools on which the software implementation will be based. Among the available programming languages, Python is the most appropriate, as it combines high performance scientific libraries, a wide range of ready-made data analysis packages, and flexible syntax that simplifies the process of code development and debugging. Python also has a large developer community that constantly maintains and improves open-source libraries for computing, machine learning, and visualisation [23–25].

The efficient implementation of neural networks and related machine learning algorithms requires the use of reliable and optimised libraries capable of performing a wide range of operations: from basic linear algebraic calculations and visualisation to advanced deep learning. This subsection considers the feasibility of using standard Python libraries (NumPy, Matplotlib, scikit-learn, SciPy, PyTorch, openpyxl, os, tkinter, time), each of which complements the other, forming an integrated ecosystem for building a high-precision information and measurement system.

This choice is driven by the need to ensure, on the one hand, sufficient performance during large-dimensional numerical computations (for example, operations with multidimensional arrays), and, on the other hand, flexibility in implementing deep network architectures and ease of developing graphical user interfaces [26].

Let's consider the feasibility of using NumPy, Matplotlib, scikit-learn, SciPy, PyTorch, openpyxl, os, and tkinter libraries to create such a system.

NumPy is a fundamental library for scientific computing in Python. It provides support for multidimensional arrays and matrices, as well as a large collection of mathematical functions for working with these arrays [27–29]. NumPy is used to generate synthetic data, perform linear algebraic operations, and calculate statistical indicators.

Compared to libraries that perform similar functions (Table 3.2), the choice of the NumPy library is reasonable in terms of high computational efficiency and efficient memory usage, which is important when processing large amounts of data.

Matplotlib is a library for creating static, animated, and interactive visualisations in Python. Compared to libraries that perform similar functions (Table 3.3), the choice of Matplotlib is a reasonable choice due to its high rendering speed and a wide range of graph types, which allows you to create a variety of visualisations of measurement results.

Scikit-learn is one of the most popular machine learning libraries for Python. Compared to libraries that perform similar functions (Table 3.4), Scikit-learn is a reasonable choice due to its learning speed and a wide range of algorithms that allow you to process data efficiently.

Table 3.2 Comparison of NumPy with analogues

Library	Matrix multiplication time (1000×1000) (c)	Memory usage (array $1\times10^{(6)}$ elements) (MB)
NumPy	0.14	8.0
Pandas	0.29	24.0
Python list	438.15	76.2

Table 3.3 Comparison of Matplotlib with analogues

Library	Rendering time (1×105 points) (s)	Number of chart types
Matplotlib	0.8	40
Seaborn	1.2	18
Plotly	1.5	30

Table 3.4 Comparison of Scikit-learn with analogues

Library	Learning time (linear regression, 1 M samples) (s)	Classification accuracy (MNIST) (%)
Scikit-learn	0.8	96.8
TensorFlow	1.5	99.2
Keras	1.7	98.9

SciPy is a library for scientific and technical computing that extends the capabilities of NumPy [30–33]. SciPy is used to: generate random numbers and calculate values for calculating confidence intervals. SciPy is a reasonable choice due to its high speed and accuracy (Table 3.5).

PyTorch is one of the leading deep learning libraries that provides GPU-enabled tensor computing and automatic differentiation. PyTorch is used to create the neural network architecture, define the loss function and optimiser, train and evaluate the model. The choice of PyTorch is based on its flexibility, dynamic computational graphs, and a large community of developers that ensures continuous improvement and support of the library (Table 3.6).

Openpyxl is a library for reading and writing Excel files in the xlsx/xlsm format. Openpyxl allows you to work with all aspects of Excel files, including styles, charts and images. The Openpyxl library is optimised for large files and has extensive formatting capabilities, which is important for creating detailed measurement reports (Table 3.7).

Tkinter is a standard Python library for creating graphical user interfaces [34] and is used to create interactive windows to display data processing results. Tkinter is a reasonable choice due to its fast startup, embeddedness in the standard Python library, and high cross-platform compatibility, which makes it easy to deploy the system on different platforms (Table 3.8).

Table 3.5 Comparison of SciPy with analogues

Library	Time to solve a system of linear equations (1000 × 1000) (s)	Accuracy of calculating integrals
SciPy	0.3	1e−8
SymPy	5.2	1e−12
MATLAB	0.4	1e−9

Table 3.6 Comparison of PyTorch with analogues

Library	Learning time (ResNet50, ImageNet, per epoch) (s)	Memory usage (ResNet50, batch_size = 32) (GB)
PyTorch	240	5.1
TensorFlow	270	5.3
MXNet	255	5.2

Table 3.7 Comparison of openpyxl with analogues

Library	Write time 1x104 lines (s)	Support for formatting	Compatible with the latest versions of Excel
Openpyxl	0.9	High	Full
Xlsxwriter	1	High	Full
Xlwt	0.7	Limited	Only xls

Table 3.8 Comparison of tkinter with analogues

Library	Simple window launches time (ms)	Built into the standard Python library	Cross-platform compatibility
Tkinter	15	Yes	Yes
PyQt	350	No	Yes
WxPython	200	No	Yes

The choice of software libraries NumPy, Matplotlib, scikit-learn, SciPy, PyTorch, openpyxl and tkinter is appropriate in terms of performance, functionality and ease of use.

These libraries provide the necessary tools for efficient data processing, visualisation of results and implementation of machine learning algorithms, which creates a powerful and flexible toolkit for the development and implementation of complex information and measurement systems. Their choice is justified by both their individual advantages and their ability to interact effectively with each other, creating an integrated ecosystem for scientific computing and data analysis.

The use of these libraries can significantly speed up the development process, ensure high system performance and create a flexible architecture that can be easily adapted to various tasks of controlling the geometric parameters of power equipment.

3.5 Conclusions

In the third chapter, the structure and functional content of the SIS with an integrated neural network for IMS based on the CMA is developed, which allows automating the processes of collecting, processing and compensating for measurement data errors.

The proposed structural diagram of the complex ensures effective interaction of all functional components, including the primary data acquisition module, the module for calculating spatial coordinates, diagnostic and calibration modules, the user interface, and information storage and transmission systems.

The choice of the architecture and parameters of the neural network, which is integrated into the measuring system in order to effectively compensate for non-kinematic errors, which significantly increases the accuracy and stability of the CMA operation, is substantiated. The developed graphical user interface provides convenient control of the complex, provides operational information about the system status, measurement and calibration results, and simplifies the operator's interaction with neural network algorithms.

Thus, the SIS make it possible to implement the proposed methods of error compensation in production conditions, which confirms the feasibility of its use to ensure an increase in the accuracy of measurements of geometric parameters of power equipment parts.

References

1. Osawa, S., et al.: Multiple orientation technique for calibration of cylindrical workpieces on CMMs. Precis. Eng. **29**(1) (2005). https://doi.org/10.1016/j.precisioneng.2004.04.006
2. Judd, R.P., et al.: Technique to calibrate industrial robots with experimental verification. IEEE Trans. Robot Automat. (1990). https://doi.org/10.1109/70.88114
3. Li, Z., Li, S., Luo, X.: A new robot arm calibration method based on cubic interpolated beetle antennae search approach. ArXiv (2022). https://doi.org/10.48550/arXiv.2204.06212
4. Alici, G., et al.: A systematic technique to estimate positioning errors for robot accuracy improvement using laser interferometry based sensing. Mech. Mach. Theory (2005). https://doi.org/10.1016/j.mechmachtheory.2004.12.012
5. Weckenmann, A., Beetz, S., Lorz, J.: Monitoring coordinate measuring machines by user-defined calibrated parts. In: Models for Computer Aided Tolerancing in Design and Manufacturing. Springer (2007). https://doi.org/10.1007/1-4020-5438-6_14
6. El Asmai, S., Hennebelle, F., Coorevits, T., et al.: Rapid and robust on-site evaluation of articulated arm coordinate measuring machine performance. Meas. Sci. Technol. (2018). https://doi.org/10.1088/1361-6501/aade10
7. Gao, G., Wang, S., Liu, F., et al.: Residual correction for robotic articulated arm coordinate measuring machine with radial basis function neural network. Int. J. Adv. Robot. Syst. (2020). https://doi.org/10.1177/1729881420925638
8. Hernandez, B.H., Plascencia, L.J., Arana, D.N., Lopez, C.F.: Inverse kinematics of robotic manipulators based on hybrid differential evolution and Jacobian pseudoinverse approach. Algorithms (2024). https://doi.org/10.3390/a17100454
9. Bachmann, J., et al.: Aide in decision-making: contribution to uncertainties in three-dimensional measurement. Precis. Eng. **28**(1) (2004). https://doi.org/10.1016/S0141-6359(03)00079-5
10. Xu, Y., Gao, Q., Limin, Y., Yong, Y.: Single-point residual correction method for articulated arm coordinate measuring machine based on pose configuration's polynomial model. In: IEEE 7th Annual International Conference on CYBER Technology in Automation (2017). https://doi.org/10.1109/CYBER.2017.8446573
11. Gao, G., Wang, S.: Residual correction for robotic articulated arm coordinate measuring machine with radial basis function neural network. Int. J. Adv. Robot. Syst. (2020). https://doi.org/10.1177/1729881420925638
12. Ostrowska, K., Gaska, A., Kupiec, R., et al.: Comparison of accuracy of virtual articulated arm coordinate measuring machine based on different metrological models. Measurement. **133** (2018). https://doi.org/10.1016/j.measurement.2018.10.024
13. Cheng, L., Wang, W., Weng, Y., et al.: A novel kinematic parameters identification method for articulated arm coordinate measuring machines using repeatability and scaling factor. Math. Probl. Eng. **2018**, 1–10 (2018). https://doi.org/10.1155/2018/1760190
14. Gao, G., Sun, G., Na, J., et al.: Structural parameter identification for 6 DOF industrial robots. Mech. Syst. Signal Process. **113**, 145–155 (2018). https://doi.org/10.1016/j.ymssp.2017.08.011
15. Luo, Z., Liu, H., Li, D., et al.: Analysis and compensation of equivalent diameter error of articulated arm coordinate measuring machine. Meas. Control. **51**, 16–26 (2018). https://doi.org/10.1177/0020294018755324
16. Feng, C.-X.J., et al.: Design and analysis of experiments in CMM measurement uncertainty study. Precis. Eng. **31**(2) (2007). https://doi.org/10.1016/j.precisioneng.2006.03.003
17. Bing, H.E., Che, L., Liu, C., et al.: DE-based algorithm for solving the inverse kinematics on a robotic arm manipulators. J. Phys. Conf. Ser. (2021). https://doi.org/10.1088/1742-6596/1922/1/012008
18. Huang, J., Ri, S., Fukuda, T., et al.: A disturbance observer based sliding mode control for a class of underactuated robotic system with mismatched uncertainties. IEEE Trans. Autom. Control. **64**, 2480–2487 (2019). https://doi.org/10.1109/TAC.2018.2868026

19. Gao, G., Zhang, H., San, H., et al.: Kinematic calibration for industrial robots using articulated arm coordinate machines. Int. J. Model. Identif. Control. (2019). https://doi.org/10.1504/IJMIC.2019.096816
20. Pulse repetition frequency multiplier: pat. 153272 Ukraine: G05D 19/00 H03K 3/00/ D. P. Ornatsky et al. No. u202202564; appl. 18.07.2022; publ. 14.06.2023
21. Wan, L.L., Su, Y.X., Zhang, H.J., et al.: Global fast terminal sliding mode control based on radial basis function neural network for course keeping of unmanned surface vehicle. Int. J. Adv. Robot. Syst. **16**, 1–12 (2019). https://doi.org/10.1177/1729881419829961
22. Park, B.J., Pedrycz, W., Oh, S.K.: Polynomial-based radial basis function neural networks (P-RBF NNs) and their application to pattern classification. Appl. Intell. **32**, 27–46 (2010). https://doi.org/10.1007/s10489-008-0133-z
23. Shaw, L., Weckenmann, A.: Automatic registration method for hybrid optical coordinate measuring technology. CIRP Ann. Manuf. Technol. **60**, 539–542 (2011). https://doi.org/10.1016/j.cirp.2011.03.055
24. Hamana, H., Tominaga, M., Ozaki, M., Furutani, R.: Calibration of articulated arm coordinate measuring machine considering measuring posture. Int. J. Autom. Technol. **5**(2) (2010). https://doi.org/10.20965/ijat.2011.p0109
25. Rak, M., Ratajczyk, E.: Coordinate measurements performed on coordinate measuring arms in manufacturing techniques. Przegląd Mechaniczny. **9** (2007). https://doi.org/10.17814/mechanik.2017.12.200
26. Mailhe, J., et al.: Geometrical checking by virtual gauge, including measurement uncertainties. CIRP Ann. Manuf. Technol. (2008). https://doi.org/10.1016/j.cirp.2008.03.112
27. Bringmann, B., et al.: A measuring artifact for true 3D machine testing and calibration. CIRP Ann. Manuf. Technol. (2005). https://doi.org/10.1016/S0007-8506(07)60147-4
28. De Chiffre, L., et al.: Comparison of coordinate measuring machines using an optomechanical hole plate CIRP Ann. Manuf. Technol. (2005). https://doi.org/10.1016/S0007-8506(07)60149-8
29. Schwenke, H., Siebert, B., Wäldele, F., Kunzmann, H.: Assessment of uncertainties in dimensional metrology by Monte Carlo simulation: proposal of a modular and visual software. CIRP Ann. Manuf. Technol. **49**, 395–398 (2000). https://doi.org/10.1016/S0007-8506(07)62973-4
30. Sladek, J., Gąska, A., Olszewska, M., Kupiec, R.: Virtual coordinate measuring machine built using Lasertracer system and spherical standard. Metrol. Meas. Syst. (2013). https://doi.org/10.2478/mms-2013-007
31. Balsamo, A., et al.: Towards instrument-oriented calibration of CMMs. Ann. CIRP. (1996). https://doi.org/10.1016/S0007-8506(07)63106-0
32. Furtani, R., Shimojima, K., Takamasu, K.: Parameter calibration for non-Cartesian CMM. VDI Ber, pp. 317–326 (2004). http://www.nanolab.t.u-tokyo.ac.jp/pdffiles/45-furutani.pdf
33. Balsamo, A., et al.: Evaluation of CMM uncertainty through Monte Carlo simulations. CIRP. (1999). https://doi.org/10.1016/S0007-8506(07)63218-1
34. Piratelli-Filho, A., Di Giacomo, B.: CMM uncertainty analysis with factorial design. Precis. Eng. **27**(3) (2003). https://doi.org/10.1016/S0141-6359(03)00035-7

Chapter 4
Algorithm for Generating and Normalising Synthetic Calibration Data for Coordinate Measuring Systems

4.1 Introduction

In the process of implementing the software and information system, one of the key aspects is the experimental verification of the effectiveness of the developed methods of compensating for non-kinematic errors in conditions as close as possible to real production conditions. Particular attention is paid to the procedure for generating controlled synthetic data sets that serve as the basis for validating the operation of the neural network integrated into the system. This approach makes it possible to formally assess the ability of the NN to identify and compensate for non-kinematic errors caused by external destabilising factors.

One of the most appropriate objects for constructing such validation scenarios is a spherical reference, since its geometry can be easily formalised analytically, and the point coordinates on its surface can be specified with high mathematical accuracy. The use of a spherical reference allows you to create a representative set of input data with clearly defined geometric characteristics, which, in turn, provides an opportunity to accurately interpret the results of error modelling. In addition, the absence of complex topological features allows us to focus exclusively on the analysis of non-kinematic effects without mixing them with geometric surface ambiguities.

In this section, we propose an algorithm for generating synthetic coordinate data sets based on the generation of random points on the surface of a spherical reference with the subsequent imposition of modelled noise. These noise components simulate typical forms of errors that occur in real-world operation of coordinate measuring systems: from phase and amplitude disturbances in the signal to thermal and mechanical instabilities in the components of the CMA. Thus, a controlled environment is created in which it is possible to isolate and study the impact of individual destabilising factors on measurement accuracy.

The results of simulation modelling allow not only to quantify the effectiveness of neural network methods of error compensation, but also to identify the limits of their application, the degree of sensitivity to the type and intensity of disturbances, as well

D. Kataiev et al., *Neural Calibration of Coordinate Measuring Arms*, Studies in Systems, Decision and Control 695, https://doi.org/10.1007/978-3-032-32918-9_4

as the ability to generalise to new data configurations. This approach significantly increases the validity of recommendations for the implementation of NN in software and information systems operating in a production environment [1–3].

4.2 Methods of Generating a Synthetic Dataset

The use of synthetic data sets is proposed to evaluate and verify the effectiveness of nonkinematic error compensation algorithms. In order to ensure the efficient operation of the software and information system and to select the optimal method for generating synthetic data sets, a number of the most common methods were analysed, their advantages and disadvantages were identified, and an algorithm was proposed (Fig. 4.1), which allows a reasonable choice of method depending on the specific conditions and objectives of the study. In addition, it is possible to add noise to simulate the influence of nonkinematic errors, which brings synthetic data closer to the real conditions of the information and measurement system. Synthetic data are generated around a 30 mm diameter calibration sphere, which meets common standards in coordinate measurements, and 1000 generated synthetic coordinates provide a balance between research accuracy and acceptable computing speed.

Let's consider the methods of synthetic data generation, their advantages and disadvantages. In most cases, the method of uniform distribution, the method of normal distribution, and the Fibonacci spiral method are used to generate synthetic data sets. An algorithm (Fig. 4.1) has been developed that allows a reasonable choice of the method for generating a synthetic data set with the subsequent application of the necessary method of adding noise to simulate the effect of nonkinematic errors.

Let us consider in more detail the methods of generating synthetic data for simulating the calibration process with a sphere.

The uniform distribution (Fig. 4.2) provides statistically uniform surface coverage (using the appropriate angular formulas), which is well suited to tasks where it is necessary to model the same probability of measurements from different areas of the sphere.

The angles φ and θ can be obtained using uniform random variables U and V in the range [0, 1]. The coordinates are then calculated using the formulas:

$$\begin{aligned}
\varphi &= \arccos(1 - 2U);\\
\theta &= 2\pi V;\\
x &= R\ \sin(\varphi)\cos(\theta);\\
y &= R\ \sin(\varphi)\sin(\theta);\\
z &= R\ \cos(\varphi),
\end{aligned} \tag{4.1}$$

where R is the radius of the sphere.

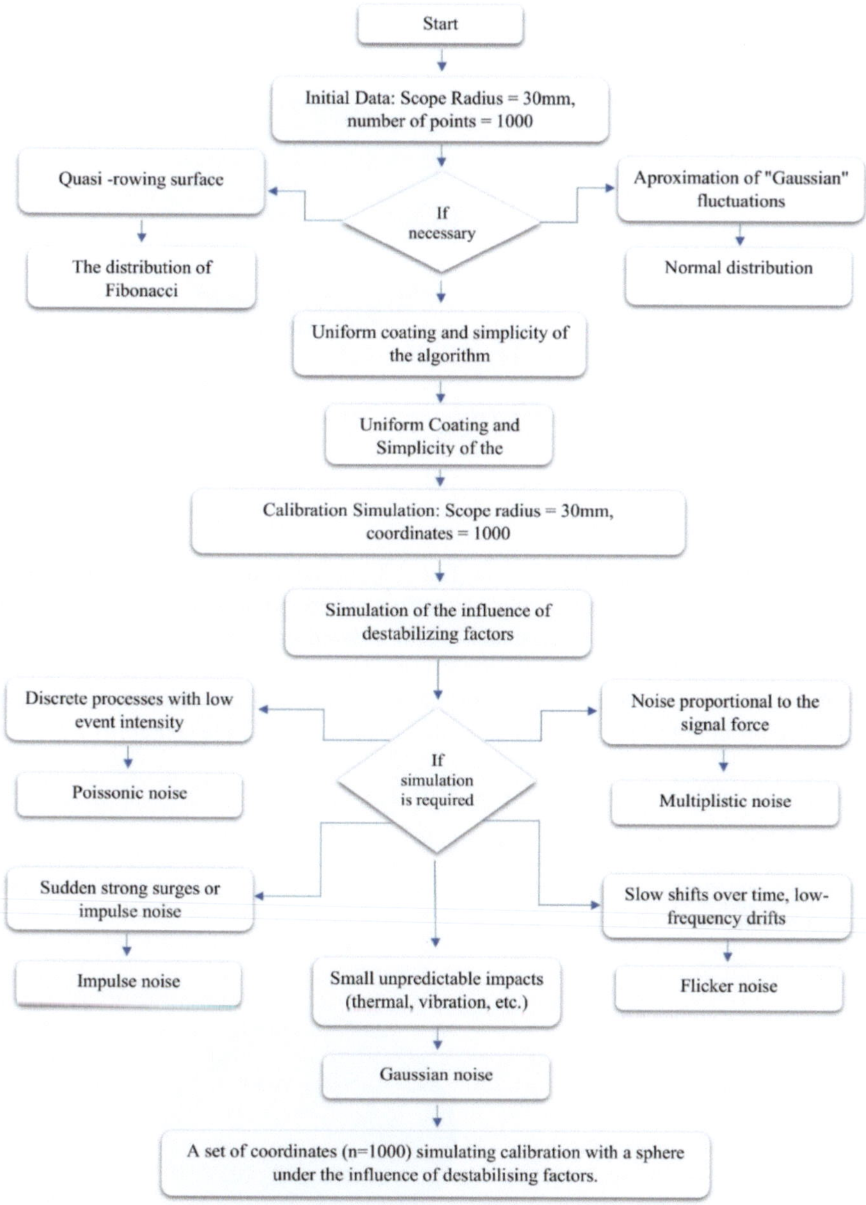

Fig. 4.1 Algorithm for selecting a method for generating a synthetic dataset

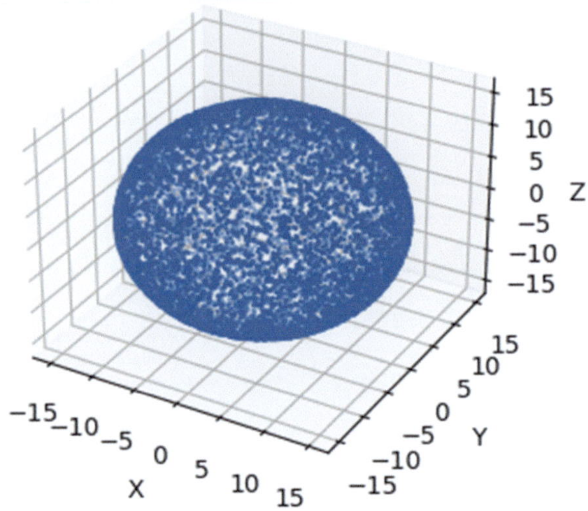

Fig. 4.2 The uniform distribution method

The normal (Gaussian) distribution (Fig. 4.3) allows to draw on the analogy with the central limit theorem and simulate situations where points can form a cloud with random deviations close to Gaussian, but still ordered on a sphere.

The normal distribution is implemented by generating three independent random variables (x, y, z) from the standard normal distribution, after which these random points are brought to the surface, i.e., scaled to the radius of the sphere.

$$(x, y, z) = N(0, 1)^3;$$
$$r = \sqrt{x^2 + y^2 + z^2};$$

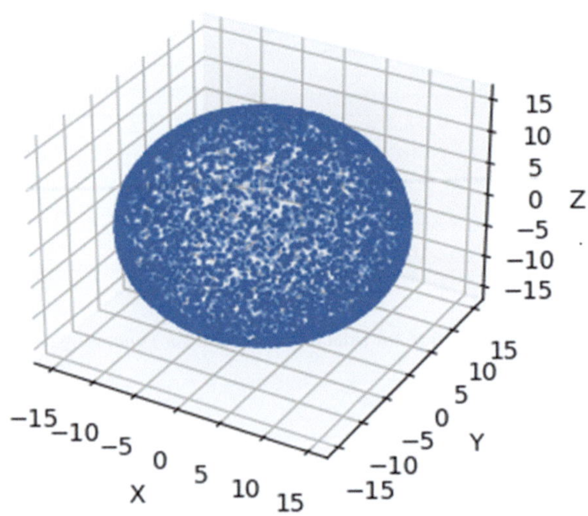

Fig. 4.3 Normal distribution method

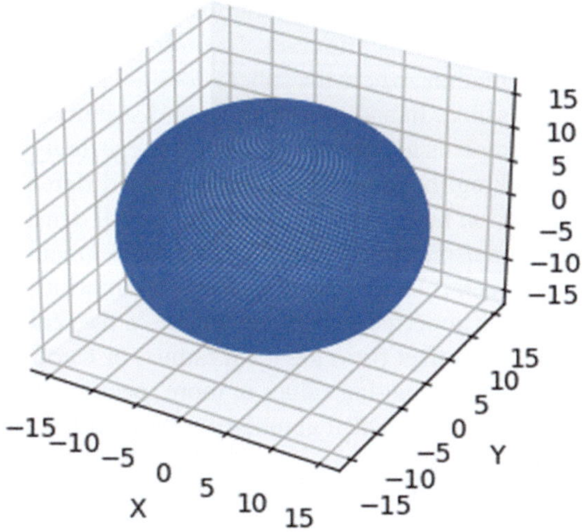

Fig. 4.4 Fibonacci spiral method

$$(x, y, z) = R\left(\frac{x}{r}, \frac{y}{r}, \frac{z}{r}\right), \tag{4.2}$$

where N (0, 1) is a standard normal distribution.

The Fibonacci spiral method (Fig. 4.4) offers one of the most efficient algorithms for quasi-uniform surface filling, avoiding the "bumps" or "gaps" that sometimes appear in pseudo-random methods.

The Fibonacci spiral method is based on the formation of a sequence of points by rotation of the angle with an incremental increase determined by the golden ratio $\pi\left(3-\sqrt{5}\right)$, for the i-th point ($i = 0, 1, \ldots, n-1$), the rotation index θ is taken equal to $F\,i$, and the coordinate z is distributed in the interval $[-1, +1]$.

$$\begin{aligned} F &= \pi\left(3-\sqrt{5}\right); \\ \theta &= F\,i, \quad i = 0, 1, \ldots, n-1; \\ z &= 1 - \frac{2i+1}{n}, \quad i = 0, 1, \ldots, n-1; \\ r &= \sqrt{1 - z2}; \\ x &= r\cos(\theta); \\ y &= r\sin(\theta), \end{aligned} \tag{4.3}$$

where n is the number of points.

The most appropriate method for simulating the calibration sphere is to use a uniform distribution, primarily because of its ease of implementation and sufficient coverage of the surface [4–6]. During the generation of points in spherical coordinates, when the azimuth angle and polar angle of φ are randomly selected, statistically uniform positions on the sphere are obtained. This method allows you to quickly obtain a set of coordinates without resorting to more complex methods such as the Fibonacci spiral. This preserves the probabilistic irregularity of the points, which is good for situations where the sphere does not necessarily need to be covered as regularly as possible. In addition, a uniform distribution is convenient for cases where a random sample is required, which can be repeatedly generated anew with a changed seed, checking the algorithm's response to different implementations. The uniformity condition helps to ensure that the overall density of points in the southern and northern hemispheres (or in different zonal segments) does not differ significantly.

The realism is formed by the additional imposition of noise, which reflects small deviations or imperfections of the sphere in the real environment. These distortions can be radial (offset from the ideal radius) or Cartesian (corrections at Δx, Δy, Δz), modelled using Gaussian values or more specific distributions. A uniform base mesh serves as the starting ideal geometry, while artificial noise adds a realistic stochastic component.

As a result, a uniform distribution of coordinates on the surface of the sphere, followed by the introduction of noise, becomes a convenient and fairly natural way to synthetically simulate the calibration environment. It does not complicate the generation, avoids potential gaps or excessive precision (as in the Fibonacci spiral method), and at the same time provides the randomness necessary to test processing algorithms. This approach is especially appropriate in tasks where the key role is played by the ability of the neural network to work out various spatial and noise scenarios with moderate homogeneity of placement [7–10].

4.3 Justification for the Method of Simulating the Impact of Destabilising Factors

The use of synthetic noise is proposed to model the residual non-kinematic error that remains after compensation of kinematic errors by traditional methods. The residual non-kinematic error arises as the total effect of numerous factors (signal distortion, temperature fluctuations, mechanical impacts, etc.) that are unique to each production environment and cannot be separated and accurately calculated by algebraic methods. Therefore, the residual non-kinematic error is considered as a single aggregate value, for the effective compensation of which it is necessary to carry out a calibration procedure in real production conditions using a neural network. Under production conditions, the system can receive additional information, for example, from auxiliary sensors (temperature, humidity, pressure) and adaptively adjust its parameters to compensate for specific environmental influences. Thus, an additional calibration procedure is carried out using a calibration sphere, in which the NN is trained to compensate for the residual non-kinematic errors inherent in this particular IMS under the relevant unique production conditions. Thus, synthetic noise simulates the general influence of factors that are difficult to approximate by classical methods, and the use of NNs in the calibration process allows for adaptive compensation of these errors in real-world operating conditions.

Each type of noise can affect the final measurement accuracy in its own way. Even minor deviations can significantly distort the results, especially when it comes to high-precision instruments. Therefore, in the process of calibration, it is important to clearly understand not only the origin of a particular noise, but also its main characteristics: statistical distribution, intensity, frequency dependence. In this subsection, we consider some of the most common methods of adding noise (Gaussian, Poisson, impulse, multiplicative and flicker), which allow you to model various destabilising factors and assess how they affect the formation of measurement errors.

Particular attention is paid to the experimental verification of each method using 3D coordinates of points on the surface of a sphere. The analysis of the influence of these noise models allows us to better understand which sources adversely affect the measurement process and how well the calibration algorithms adapt to different conditions. This is especially important in applications where measurement accuracy and reliability are critical, and different types of noise can distort the original data in different ways.

Below is an overview of the five most common types of noise that are usually considered in the context of metrology and information and measurement systems. For each of them, the characteristic features, mechanisms of occurrence and influence on the process of calibration of three-dimensional coordinates are indicated.

Gaussian noise (Fig. 4.5) is the most common variant when random deviations are described by a normal distribution.

Its popularity can be explained by the central limit theorem, according to which the sum of a large number of independent factors tends to follow a normal law.

In measuring systems, these factors can include small vibrations in electronics, mechanical vibrations, and micro-deformations in joints. The addition formula looks like this:

Fig. 4.5 Gaussian noise

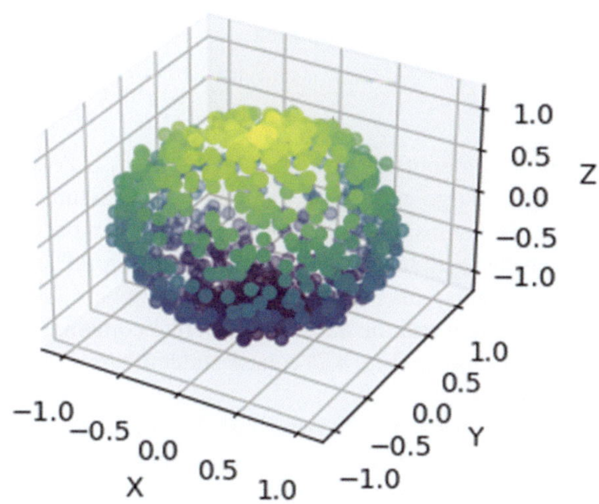

Fig. 4.6 Poisson noise

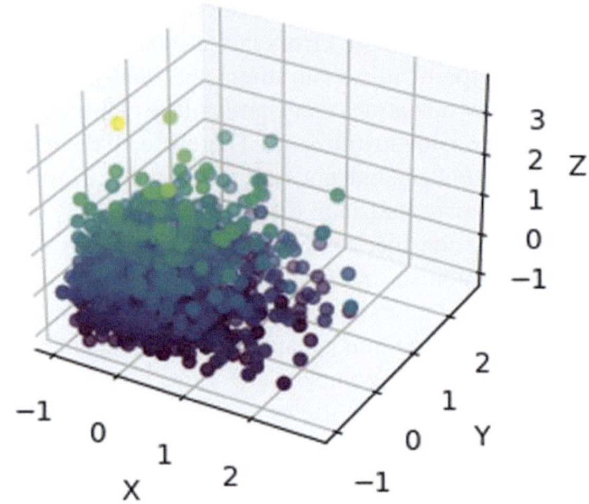

$$X_n = X + N\left(\mu, \sigma^2\right), \tag{4.4}$$

where X is the original signal, $N(\mu, \sigma^2)$ is a Gaussian random variable with mean μ and variance σ^2. Depending on the value of σ, the noise can be small (a subtle disturbance) or large enough to significantly distort the output. In most real-world devices, the total error is generally close to normal, so Gaussian noise is often considered the "most realistic" in the general case. It is also easy to control by varying μ and σ and simple to implement in code.

Poisson noise (Fig. 4.6) occurs in tasks where events are counted discretely and the intensity of these events is proportional to the input signal.

Examples include photoelectronic processes (in particular, photon counters), radiometric measurements of small signals, dosimeters, etc. Poisson noise describes the probability that a certain number of events will be "detected" during a certain time interval or in a certain area:

$$X_n = P(\lambda X), \tag{4.5}$$

where $P(\lambda)$ denotes a random variable with a Poisson distribution and parameter λ. If λX is large enough, it can be roughly assumed to be close to Gaussian. But where the intensity of the process or the number of events is small, the Poisson noise fully reproduces the character of "threshold" or "quantised" oscillations. In 3D measurements, this can model conditions where the number of "pulses" (e.g., laser glare) is small and the data has a discrete component.

Impulse noise (Fig. 4.7) causes sudden bursts of values, often far beyond normal fluctuations.

Such pulses can be the result of short-term electromagnetic interference, a sharp impact, exceeding the sensor's sensitivity threshold, or an electronics failure. In coordinate measurements, this is sometimes observed when the instrument "jerks"

Fig. 4.7 Impulse noise

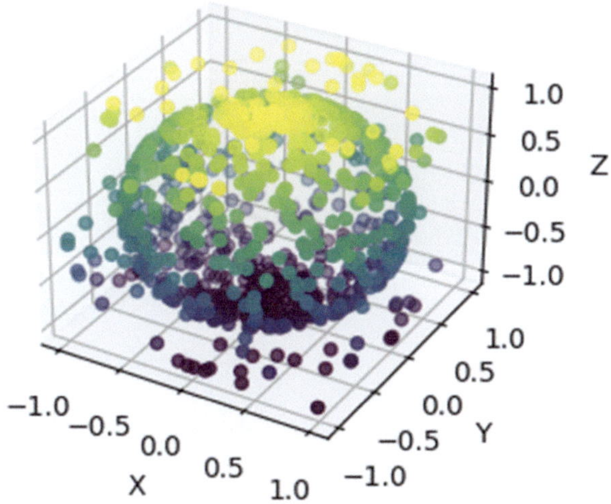

or produces inaccurate data outputs at short times. In formal terms, impulse noise is described as randomly spaced "peaks" whose values are significantly higher than the typical deviation:

$$X_n = \begin{cases} X \\ X + \Delta \end{cases}. \tag{4.6}$$

Such "spikes" can potentially cause large systematic errors if the calibration system does not take them into account or reject them. Impulse noise is most often simulated to test the stability of an algorithm: whether it is able to filter out short-term anomalies.

Multiplicative noise (Fig. 4.8) is noising whose intensity depends on the signal value.

Unlike additive models, where a random variable is simply added to the signal, this type of noise implies that the intensity depends on the amplitude of the signal itself:

$$X_n = X^* \left(1 + N\left(0, \sigma^2\right)\right). \tag{4.7}$$

Such processes occur, for example, in radar or optical systems (laser rangefinders), where the strength of the reflected signal affects the level of interference. In 3D coordinate problems, multiplicative noise is important when more distant points are recorded with a worse signal-to-noise ratio, and closer points with a better one. This noise allows you to check whether the algorithm is "lost" in areas of high or low amplitude.

Flicker noise (Fig. 4.9) or 1/f-noise is noise whose power spectral density is inversely proportional to frequency.

Fig. 4.8 Multiplicative noise

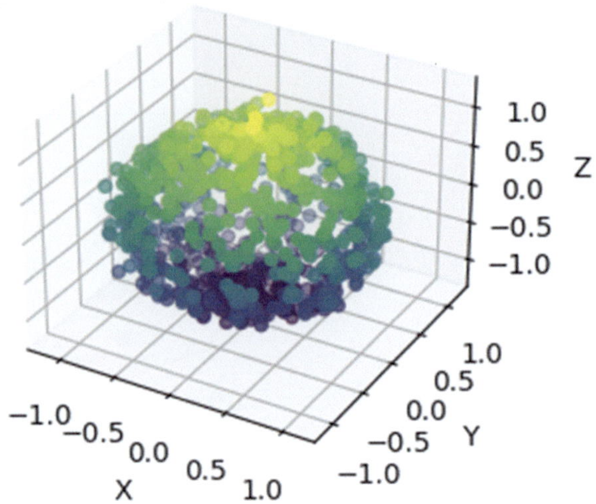

$$S(f) \propto \frac{1}{f^{\alpha}}; \quad 0 < \alpha < 2. \tag{4.8}$$

This noise is found in electronic components, transistors, lasers, and sometimes in high-precision sensors. It is especially noticeable at low frequencies, causing the device to "drift" over time. In the case of 3D coordinate measurements, this can mean a slow but stubborn change in the zero point or a gradual shift in the scale, which together reflects long-term non-stationarity. Flicker noise is more difficult to filter using classical methods, and its model is used when it is necessary to check the system's ability to track slow drifts [11–14].

Fig. 4.9 Flicker noise

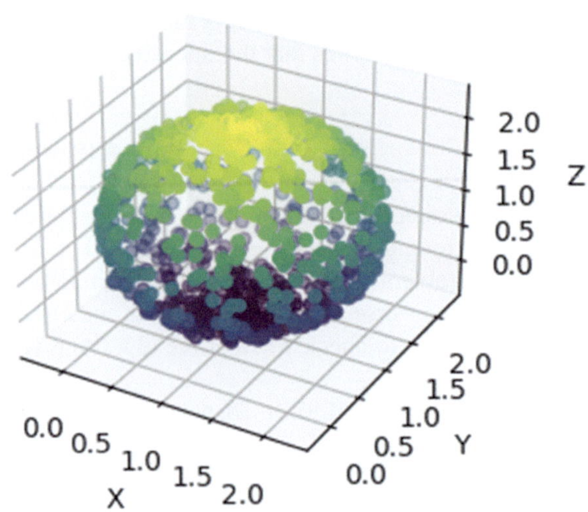

Each of these types of noise reproduces a certain way of distorting the original data, and thus can be useful for testing calibration algorithms for information and measurement systems. If the majority of error sources in real conditions produce a fairly "continuous" set of minor deviations, Gaussian noise is the most appropriate (which is often confirmed in practice).

The Gaussian (normal) distribution is fundamental in probability theory and statistics. According to the Central Limit Theorem, the sum of a large number of independent random variables tends to be normally distributed. In real-world measurement systems, many error sources have characteristics close to the normal distribution. The Gaussian distribution has well-studied mathematical properties that allow for the effective application of various methods of statistical analysis and signal processing. In many real-world systems, the total measurement error approaches a normal distribution due to a combination of different error sources. This makes Gaussian noise the most realistic model for many practical applications. Gaussian noise can be easily tuned by changing the distribution parameters (mean and standard deviation), which allows you to model different noise levels and characteristics [15–18].

Thus, among all the considered types of noise, it is the Gaussian distribution that in most cases is the most suitable for modelling the influence of destabilising factors. The versatility and correspondence to the physical nature of measurement processes make Gaussian noise the main basic choice for simulating nonkinematic errors in the tasks of calibration of information and measurement systems.

4.4 Justification of the Method of Normalisation of Input Data for a Neural Network

Formation of input and output data is an important stage in the development of an effective neural network model for information and measurement systems. This process has a significant impact on the quality of model training, its ability to generalise, and, as a result, on the accuracy of calibration. Data normalisation is the transformation of the original numerical features to a common scale in order to avoid dominance of individual variables in the process of training neural networks and reduce the negative impact of significantly different ranges. If the data is of a different nature or contains outliers, the presence of incommensurate values can make convergence difficult and cause incorrect weighting [19–22].

The min-max method (Fig. 4.10) of data normalisation involves bringing all input variables to a single scale, most often to the range [0, 1] or [−1, 1].

$$x_n = \frac{(x - x_{min})}{(x_{max} - x_{min})}, \tag{4.9}$$

where x is the initial value, x_{min} and x_{max} are the minimum and maximum values of the feature.

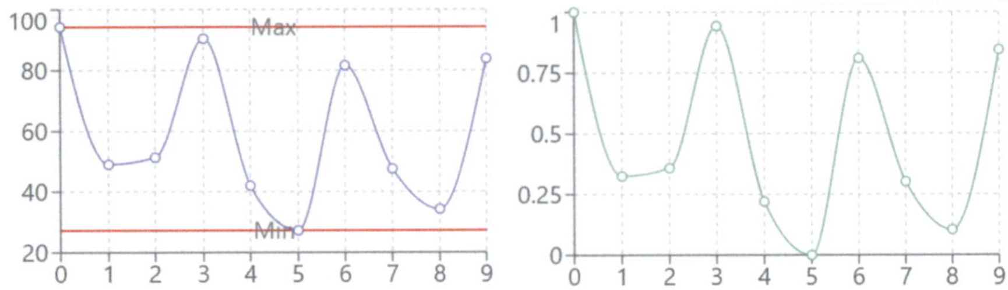

Fig. 4.10 Min-max normalisation

As a result, each dimension falls within the range [0...1], which helps to simplify the calculation and speed of implementation. However, one or more extreme outliers can significantly change the scale, forcing the remaining points to be concentrated near the centre.

Z-normalisation (or standardisation) (Fig. 4.11) brings all numerical features to zero mean and one standard deviation by the formula

$$z = \frac{(x - \mu)}{\sigma}, \tag{4.10}$$

where μ is the mean value, σ is the standard deviation.

The advantage is that the sensitivity to adding new points is reduced, as the centre remains near zero and the scatter is guaranteed to be close to one. This property is optimal for cases where the distribution of the input data is close to normal, as it increases the stability and speed of training. However, if the distribution is significantly different from the Gaussian distribution and there are noticeable outliers, the mean and standard deviation may not be representative.

Robust normalisation (Fig. 4.12) uses the median and interquartile range to remove the influence of single outliers.

In this case, the adjustment can be recorded as:

$$x_r = \frac{(x - m(x))}{IQR(x)}, \tag{4.11}$$

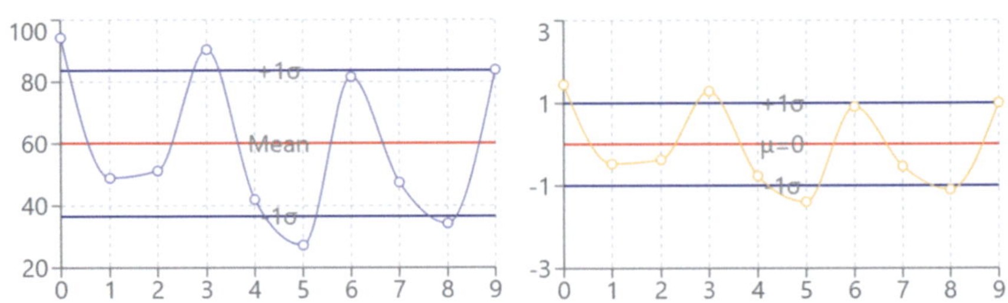

Fig. 4.11 Z-normalisation

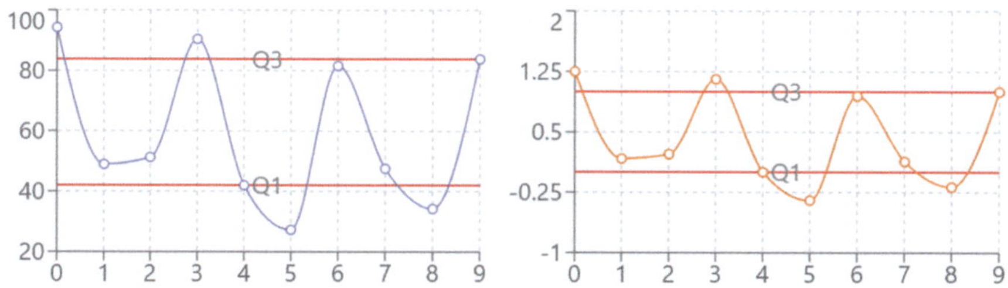

Fig. 4.12 Robust normalisation

where $m(x)$ is the median and $IQR(x)$ is the interquartile range.

This method preserves the general shape of the distribution and is less dependent on rare outliers. However, the median and IQR calculation can be more time-consuming than min-max or Z-normalisation, and does not always provide a good fit with algorithms that assume a Gaussian nature of the data.

In general, the choice of method depends on the nature of the input features and the noise level. If a full-fledged relative scale between maximum and minimum is desired, the min-max method is appropriate. If the majority of the data has a close-to-normal distribution and consistency with machine algorithms is required, Z-normalisation is preferable. If the average emissions are high and there is a risk that some points are very far from the main array, robust normalisation will allow you to focus on "typical" values without distorting the overall estimate [23–25].

Z-normalisation (standardisation) is the most appropriate method for preparing neural network input data. It allows you to bring the input characteristics, in particular the coordinates (X, Y, Z), to a single range with a zero mean and a standard deviation equal to one. This helps to increase the stability and accelerate the convergence of the neural network, as the scale of variations becomes predictable, and the impact of abnormal or excessively large values is reduced.

In the task of controlling spatial coordinates, it is important that the introduction of non-kinematic errors does not lead to uneven weighting of individual features. After standardisation, all numerical data retains a more natural structure than in the case of min-max normalisation, while remaining robust to the heterogeneous nature of noise, as it has a single mean and variance. As a result, Z-normalisation strikes an optimal balance between ease of implementation, preservation of the probabilistic distribution of coordinate data, and adaptability to noise levels, making it the best choice for developing a calibration model.

4.5 Developing an Algorithm for Distributing Input Data for a Neural Network

Building a neural network requires a correct procedure for dividing the data into training, validation and test sets, which makes it possible to find out how well the model is able to generalise the knowledge gained and whether overtraining occurs when working with experimental or real data. Depending on the type of data, we use methods of random data distribution, K-fold cross-validation, time distribution, and stratified distribution.

An algorithm has been developed to determine the method of synthetic data distribution (Fig. 4.13).

Let's take a closer look at the proposed methods.

The random distribution (Fig. 4.14) is carried out by the following formulas:

$$T_{train} = n\left(1 - T_{test_{size}}\right); \\ T_{test} = n\, T_{test_{size}}, \tag{4.12}$$

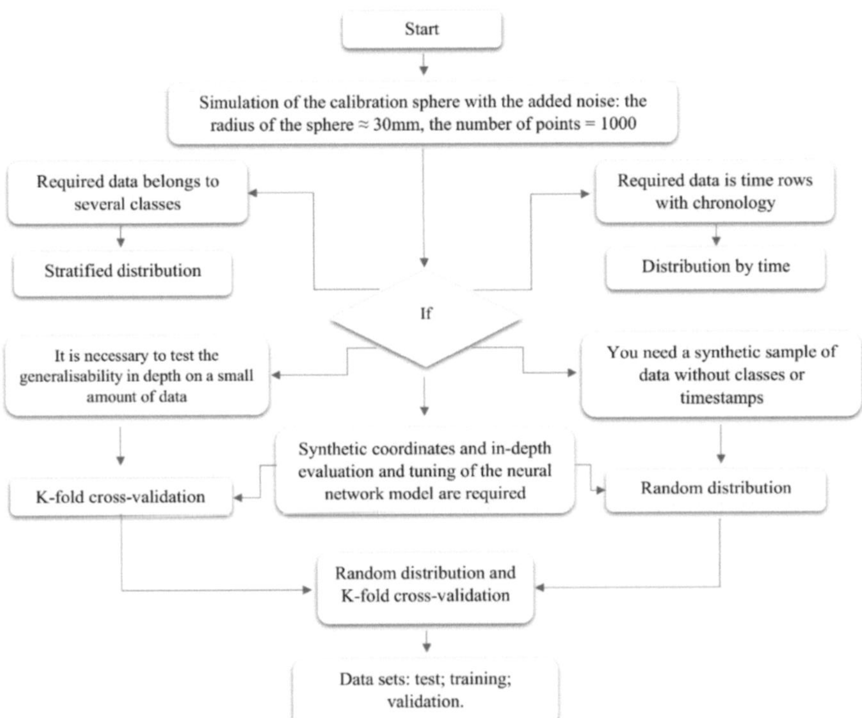

Fig. 4.13 Algorithm for determining the method of synthetic data distribution

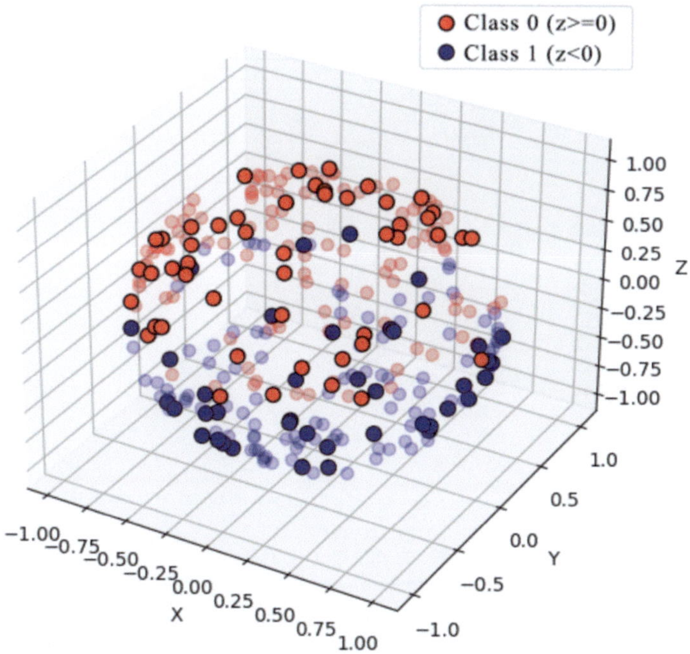

Fig. 4.14 Random distribution

where n is the total number of samples, $T_{test_{size}}$ is the proportion of the test set (usually 0.2–0.3).

In this setting, all data is shuffled and then randomly divided between the training and test subsets. The advantage of randomisation is that it does not require any additional assumptions about the temporal structure or classification, and it distributes the data evenly if it is generated according to the same laws.

K-fold cross-validation (Fig. 4.15) allows to effectively use the same data for training and validation when you need to assess the generalisability of the model in more depth.

The training set is divided into K approximately equal parts of size:

$$m = \frac{n}{K}. \tag{4.13}$$

At each iteration, one part is taken as validation data, while the other $K - 1$ parts are used as training data. The resulting error value is averaged over the number of all iterations, and this yields an objective metric known as the cross-validation error, which is usually written:

$$CV(f) = \frac{1}{K}\sum_{K=1}^{K} L(y[k], f_{-k}), \tag{4.14}$$

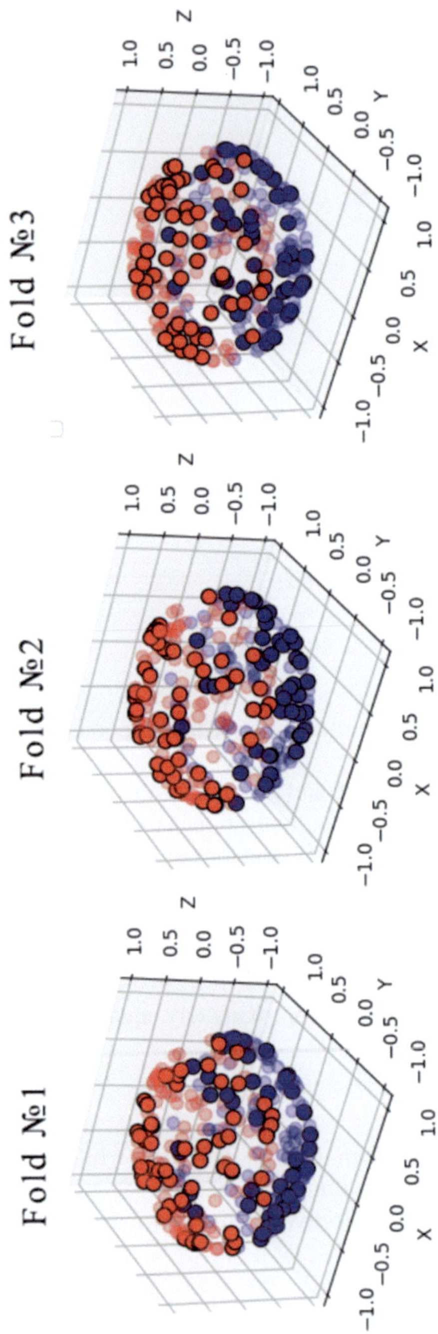

Fig. 4.15 K-fold cross-validation

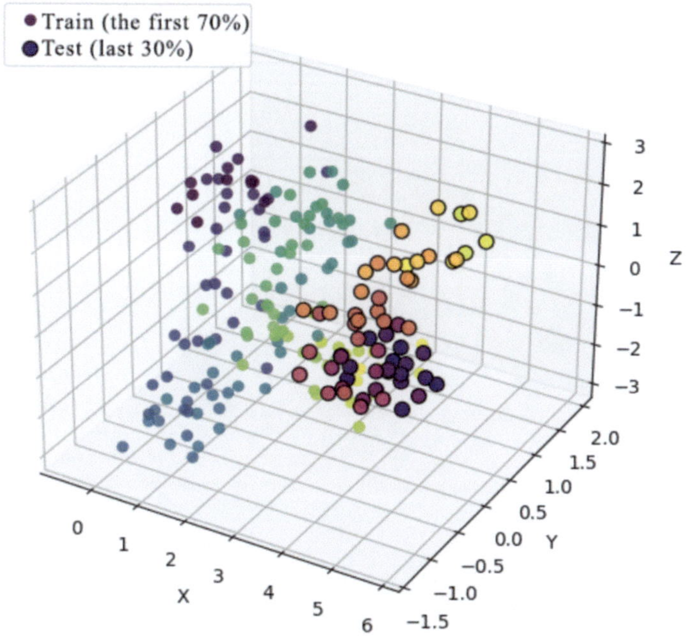

Fig. 4.16 Distribution by time

where f is the model to be validated, L is the loss function, $y[k]$ is the true values for the k block, and f_{-k} is the model trained on the rest of the sample without the k block. This cross-validation allows for a better assessment of how robust the network behaves under different distributions of training examples, which is especially useful when the data is relatively small or when selecting hyperparameters.

The distribution by time (Fig. 4.16) is appropriate when the data are time series that are chronologically ordered.

Suppose that we have ordered observations (x_i, t_i), where t_i is a timestamp (or just a monotonically increasing index) and i varies from 1 to n. Then for some threshold T_c is defined:

$$T_{train} = \{(x_i, t_i)t_i \leq T_c\}, \quad T_{test} = \{(x_i, t_i)t_i > T_c\}. \tag{4.15}$$

The distribution performed in chronological order is training, and the data after it is a test. In synthetic sphere point generation, there is no such sequence, so the time-based distribution does not provide additional benefits.

Stratified distribution (Fig. 4.17) is mostly appropriate in classification tasks with several classes or at least different subgroups with an uneven number of examples, where the proportions of each class should be proportionally preserved.

$$P\left(y = k \mid D = T_{train_{size}}\right) = P\left(y = k \mid D = T_{test_{size}}\right) = P(y = k), \tag{4.16}$$

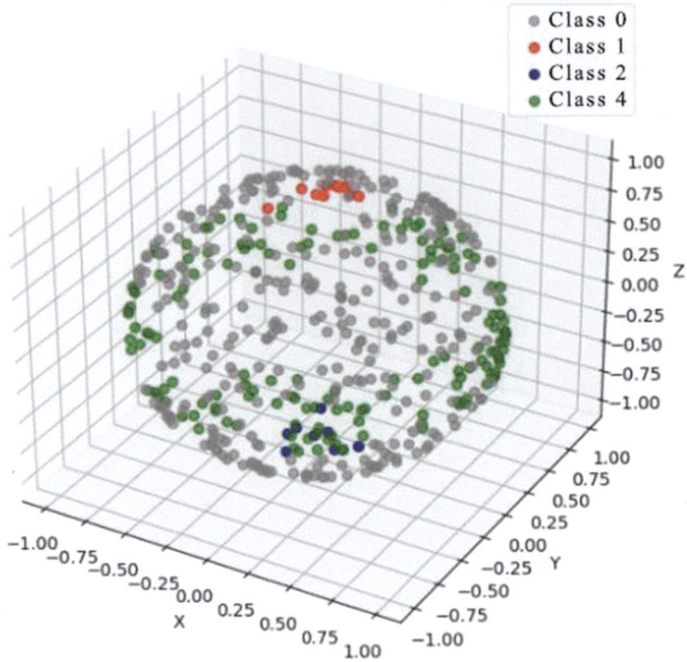

Fig. 4.17 Stratified distribution

where y is the target variable, k is the class, and D is the data set.

Synthetic coordinates generated randomly by the same distribution law do not have distinguishable classes or groups. Accordingly, for simulation tasks of the calibration procedure, the stratified distribution is of low efficiency [25–27].

Taking this into account, it was proved that the most appropriate approach is to choose a combined approach that combines random data distribution and the K-fold cross-validation procedure (Fig. 4.18). First, you can select the test set randomly, and then divide the remaining data for training into several folds.

This will allow us to accurately measure how the model performs in each fold, while at the same time providing independent verification of the results on a randomly generated test subset that is not used in the parameter selection or hyperparameter matching process. The use of random separation is due to the fact that synthetic data is generated independently of any external conditions and does not require any time intervals to be stored or taken into account. The artificially added noise in each example also has an identical distribution, so shuffling and random distribution will not lead to loss of information or distortion of statistics. For its part, K-fold cross-validation allows us to assess whether the model is not prone to overfitting when the training set configuration is changed and whether adequate accuracy is maintained when new points are substituted.

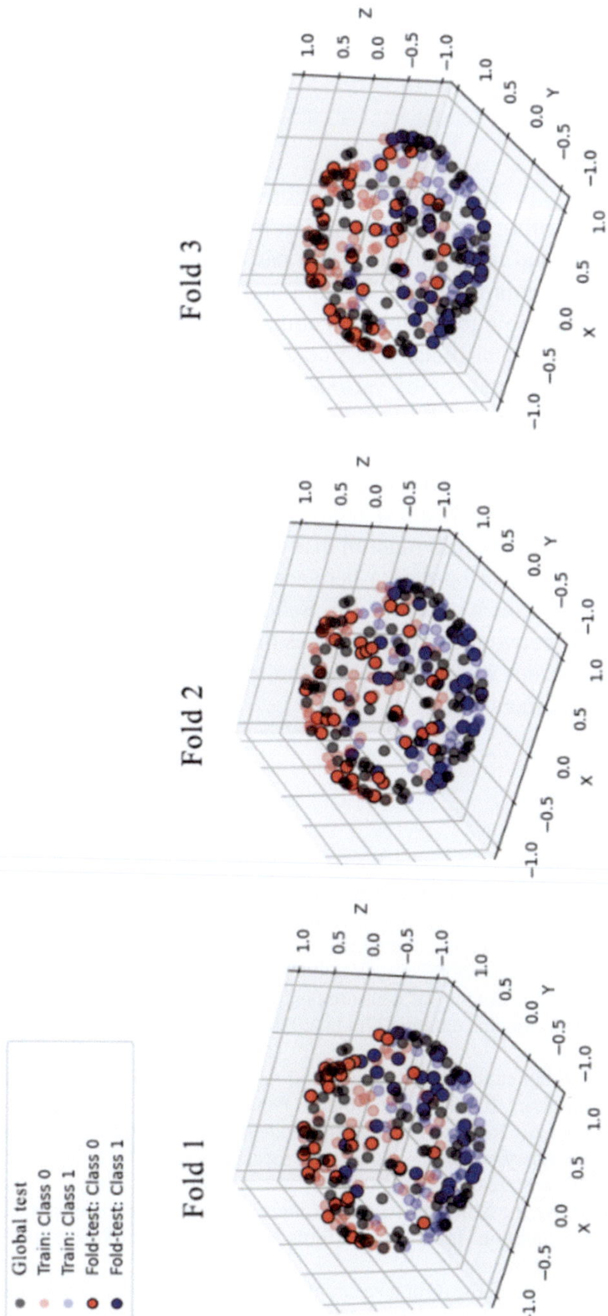

Fig. 4.18 Combined approach

This combined approach provides a thorough test of the neural network's ability to compensate for noise distortions, and also allows us to confidently determine how stable the system will be on new data that was not part of the training process.

4.6 Conclusions

A comprehensive algorithm for data preparation for training a neural network integrated into a software and information system has been developed. The method of synthetic generation of coordinates based on a uniform distribution was substantiated, which made it possible to model the geometry of the calibration sphere with the subsequent addition of artificial noise components that simulate the effect of nonkinematic errors. An analysis of various methods of simulating destabilising factors was carried out, which allowed us to determine which of them are most suitable for simulating real measurement conditions and for testing the stability of calibration algorithms. A method for normalising the input data for a neural network is proposed, which ensures scale consistency and improves model convergence. A combined approach that combines random distribution with K-fold cross-validation is proposed to correctly distribute data between training, validation and test sets. Thus, the integrated approach developed in Sect. 4.4 improves the quality of data preparation for neural network training, reduces the risk of overtraining and ensures the stability of the system when simulating the impact of nonkinematic errors in high-precision measurements.

References

1. Hyun, J.J., Hyun, K.S., Keun, Y.: Calibration of geometric and nongeometric errors of an industrial robot. Robotica **19**, 311–321 (2001). https://doi.org/10.1017/S0263574700002976
2. Besl, P.J., Ristic, M.: A method for the registration of 3D shapes. IEEE Trans. Pattern Anal. Mach. Intell. **14**(2), 239–256 (1992). https://doi.org/10.1109/34.121791
3. Muelaner, J.E., et al.: Uncertainty of measurement for large product verification: evaluation of large aero gas turbine engine datums. **27**(11) (2016). https://doi.org/10.1088/0957-0233/27/11/115003
4. Duelen, G., et al.: Robot calibration—method and results. J. Robot. Comput. Integr. Manuf. **8**(4) (1991). https://doi.org/10.1016/0736-5845(91)90034-P
5. Kruth, J.P., Kerstens, A.: Reverse engineering modelling of free-form surfaces from point clouds subject to boundary conditions. J. Mater. Process. Technol. **76**(1–3), 120–127 (1998). https://doi.org/10.1016/S0924-0136(97)00341-5
6. Gong, C., Yuan, J., Ni, J.: Nongeometric error identification and compensation for robotic system by inverse calibration. Int. J. Mach. Tools Manuf. **40**(14) (2000). https://doi.org/10.1016/S0890-6955(00)00023-7
7. Sartori, S., Zhang, G.: Geometric error measurement and compensation of machines-an update. CIRP Ann. **57**(2), 660–675 (2008). https://doi.org/10.1016/j.cirp.2008.09.008

8. Xu, X., Zhu, D., Zhang, H., et al.: TCP-based calibration in robot-assisted belt grinding of aero-engine blades using scanner measurements. Int. J. Adv. Manuf. Technol. **90**, 635–647 (2017). https://doi.org/10.1007/s00170-016-9331-8
9. Hocken, R., Simpson, J., Borchardt, B., Lazar, J., Reeve, C., Stein, P.: Three dimensional metrology. Ann. CIRP **26**(2), 403–408 (1977)
10. Kosarevsky, S.: Practical way to measure large-scale 2D parts using repositioning on coordinate-measuring machines. Measurement (2010). https://doi.org/10.1016/j.measurement.2010.02.015
11. ASME B89.4.1b: Methods for Performance Evaluation of Coordinate Measuring Machines, with Addenda. An American National Standard (2001)
12. Bräuer-Burchardt, C., et al.: High-speed three-dimensional measurements with a fringe projection-based optical sensor. Opt. Eng. **53** (2014). https://doi.org/10.1117/1.OE.53.11.112213
13. Larsson, S., Kjellander, J.A.P.: An industrial robot and a laser scanner as a flexible solution towards an automatic system for reverse engineering of unknown objects. In: Proceedings of the ASME Biennial Conference on Engineering Systems Design and Analysis, Manchester, UK, 19–22 July 2004, pp. 341–350. https://doi.org/10.1115/ESDA2004-58277
14. Saito, K., et al.: Noncontact 3-D digitizing and machining system for free-form surfaces. CIRP Ann. Manuf. Technol. **40**, 483–486 (1991). https://doi.org/10.1016/S0007-8506(07)62035-6
15. Jin, Z., et al.: Configuration analysis of the ers points in large-volume metrology system. Sensors **15**(9) (2015). https://doi.org/10.3390/s150924397
16. Mendikute, A., et al.: Self-calibrated in-process photogrammetry for large raw part measurement and alignment before machining. Sensors **17**(9) (2017). https://doi.org/10.3390/s17092066
17. Gao, G., Zhang, H., San, H., Wu, X., et al.: Modeling and error compensation of robotic articulated arm coordinate measuring machines using BP neural network. Complexity (2017). https://doi.org/10.1155/2017/5156264
18. Ye, S.H., et al.: Robot calibration using iteration and differential kinematics. J. Phys. Conf. Ser. **48**, 1–6 (2006). https://doi.org/10.1088/1742-6596/48/1/001
19. Moona, G.C., Kumar, V., Jewariya, M., Kumar, H.: Measurement uncertainty assessment of articulated arm coordinate measuring machine for length measurement errors using Monte Carlo simulation. Int. J. Adv. Manuf. Technol. **119**(2), 1–14 (2022). https://doi.org/10.1007/s00170-021-08416-1
20. Du, H., et al.: Development and verification of a novel robot-integrated fringe projection 3D scanning system for large-scale metrology. Sensors **17**(12) (2017). https://doi.org/10.3390/s17122886
21. Calvo, R., D'amato, R., Gomez, E., Domingo, R.: Integration of error compensation of coordinate measuring machines into feature measurement: Part I—model development. Sensors **16**(10), 1610 (2016). https://doi.org/10.3390/s16101610
22. Yin, S., et al.: Surface form inspection with contact coordinate measurement: a review. Int. J. Extreme Manuf. **5**(2). https://doi.org/10.1088/2631-7990/acc76e
23. Yin, S., et al.: Development and calibration of an integrated 3D scanning system for high-accuracy large-scale metrology. Measurement **54**, 65–76 (2014). https://doi.org/10.1016/j.measurement.2014.04.009
24. Limaiem, A., et al.: Integrated accessibility analysis and measurement operations sequencing for CMMs. J. Manuf. Syst. **19**(2) (2000). https://doi.org/10.1016/S0278-6125(00)80002-3
25. Shen, Y.-L., et al.: Pretravel compensation for horizontally oriented touch trigger probes with straight styli. J. Manuf. Syst. (1999). https://doi.org/10.1016/S0278-6125(99)80030-2
26. Ostrowska, K., Gąska, A., Kupiec, R., Sładek, J.: Accuracy assessment of coordinate measuring arms using laser tracer system. In: 11th International Symposium on Measurement and Quality Control (2013)
27. Liu, Z., Peng, K., Han, L.: Modeling and control of robotic manipulators based on artificial neural networks: a review. J. Sci. Technol. Trans. Mech. Eng. (2023). https://doi.org/10.1007/s40997-023-00596-3

Chapter 5
Neural Network Calibration of Coordinate Measuring Systems: Architecture, Hyperparameter Study, and Experimental Validation

5.1 Introduction

This chapter describes the process of constructing a neural network (defining its architecture, choosing an activation function and hyperparameters), as well as methods of training it. The neural network will be trained using a synthetic data set simulating the coordinates of points (the number of point coordinates is 1000) on the surface of a sphere with a diameter of 30 mm with added noise simulating the influence of destabilising factors. The choice of optimal learning parameters (learning rate, number of epochs, regularisation method) and loss functions that best reflect the target accuracy metric is experimentally substantiated. Experimental studies have been carried out to evaluate the effectiveness of the proposed software and information system (SIS) with an integrated neural network.

5.2 Experimental Justification of the Choice of the Activation Function of the Neural Network

Activation functions determine how well a neural network can approximate complex relationships between input and predicted data. In general, each layer of the network first calculates a linear combination of weights and offsets, after which the resulting value is subjected to a nonlinear transformation using an activation function. It is thanks to these functions that the model acquires the ability to learn non-trivial relationships between input and output. Typically, neural networks use activation functions such as Rectified Linear Unit (ReLU), Sigmoid, Hyperbolic Tangent (Tanh), and Leaky Rectified Linear Unit (Leaky ReLU) [1–3].

ReLU is the most common activation function (Fig. 5.1) used to introduce nonlinearity into the model:

D. Kataiev et al., *Neural Calibration of Coordinate Measuring Arms*, Studies in Systems, Decision and Control 695, https://doi.org/10.1007/978-3-032-32918-9_5

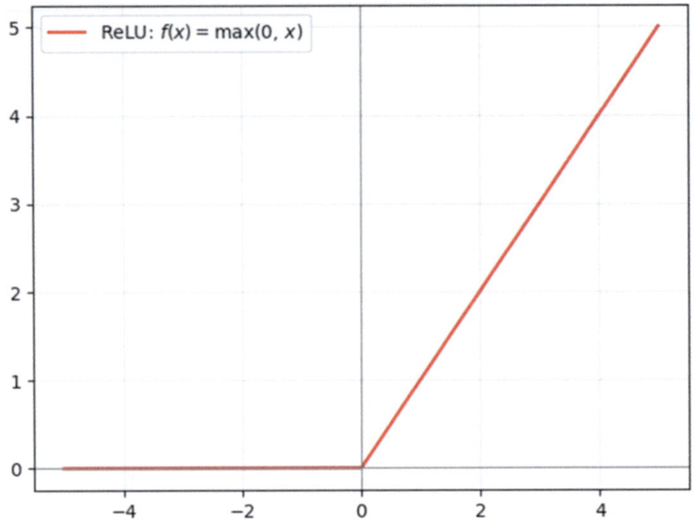

Fig. 5.1 ReLU activation function

$$f(x) = \max(0, x), \tag{5.1}$$

where x is a linear combination of weights.

The advantage of ReLU is the simplicity of computation and the ability to avoid the problem of vanishing gradients in large depth architectures, since the derivative of x for $x > 0$ is a constant. The disadvantage is the complete deactivation of neurons when the input values are negative.

For tasks where the output of a neural network requires probabilities or signals limited in the interval from zero to one, the Sigmoid function is used:

$$\mathrm{f(x)} = \frac{1}{1 + e^{(-\mathrm{x})}}. \tag{5.2}$$

Sigmoid (Fig. 5.2) is conveniently used for classification tasks where the output signal should be between 0 and 1, but the difficulty lies in the gradual saturation in the saturated part of the curve and the loss of the gradient, which slows down the learning process or leads to difficulties in optimisation.

The disadvantage of the Sigmoid function is problems with the disappearance of gradients and slow convergence.

An alternative option is the hyperbolic tangent (Tanh) (Fig. 5.3).

This function, compared to Sigmoid, provides values in the range from −1 to 1, which makes it possible to work with more balanced nonlinearity:

$$f(x) = \frac{\left(e^x - e^{(-x)}\right)}{\left(e^x + e^{(-x)}\right)}. \tag{5.3}$$

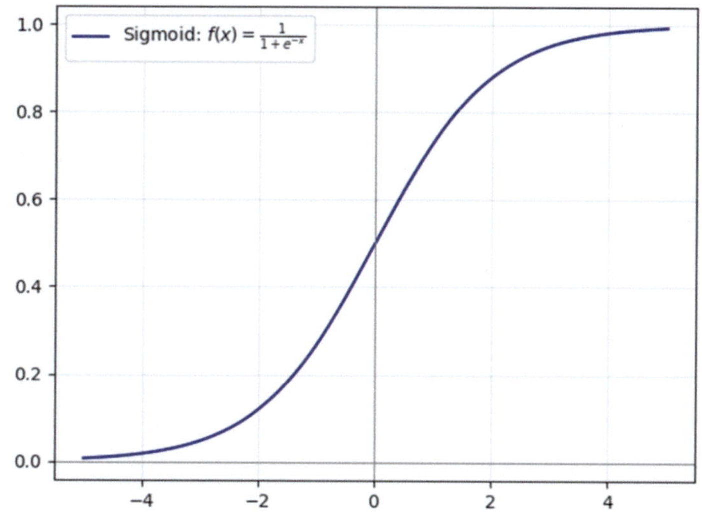

Fig. 5.2 Sigmoid activation function

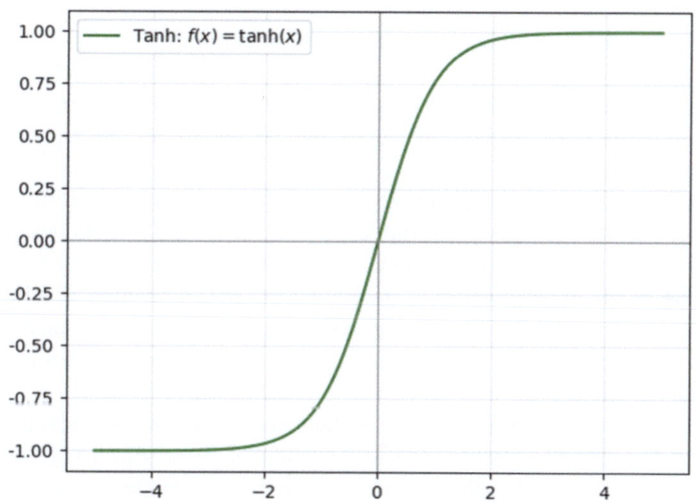

Fig. 5.3 Tanh activation function

However, it may also experience the disappearance of gradients in saturation regions where the derivative approaches zero.

Leaky ReLU (Fig. 5.4)—a modification of ReLU that solves the problem of "neuronal extinction" by allowing a small negative gradient.

$$f(x) = \max(\alpha x, x), \tag{5.4}$$

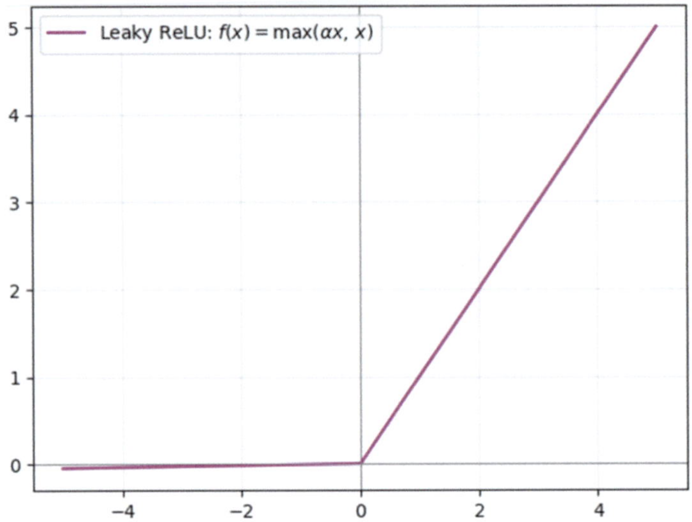

Fig. 5.4 Leaky ReLU activation function

where α is a small positive constant (usually of the order of 0.01).

Leaky ReLU helps neurons with negative inputs not to “switch off” completely, allows them to maintain a non-zero gradient, and avoids the problem of too many neurons becoming inactive, but introduces another parameter that should be tuned to achieve the best result.

An experimental study was conducted to determine the impact of different activation functions on the accuracy of the model used for CMA calibration. Four activation functions were chosen as the object of study: Sigmoid, Tanh, Leaky ReLU and ReLU. The purpose of the experiment is to determine the best activation function for the CMA calibration model by conducting 100 training iterations for each activation function. The results will be evaluated based on the average test losses obtained during each iteration.

A synthetic dataset consisting of three-dimensional coordinates was used for the experiment. The data was generated around a sphere of 30 mm diameter. The data was generated using a normal distribution, after which the coordinates were normalised to the radius of the sphere. Noise was added to these coordinates to represent real-world distortions that can occur in a measurement.

The generated data was split into training and test sets in the ratio of 80% to 20%. The training set was used to train the model, and the test set was used to evaluate its accuracy.

The architecture of the model consisted of two linear layers with a hidden layer containing 64 neurons and an output layer with 3 neurons. The model was implemented using the PyTorch library.

A separate neural network model was created for each of the four activation functions. The training process lasted 100 epochs for each iteration. The Adam

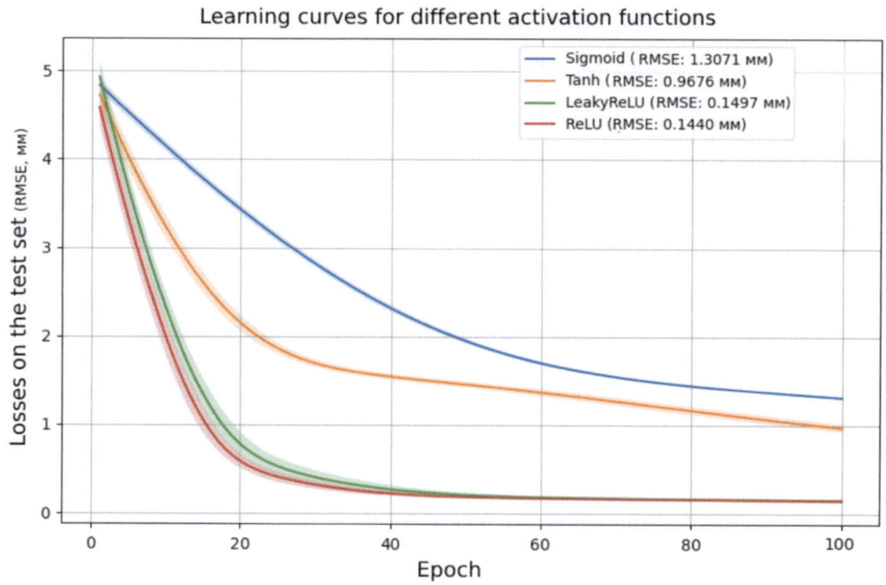

Fig. 5.5 Test losses for activation functions

algorithm with a learning rate of 0.001 was used to optimise the weights. The mean square error (MSE) was used as a loss function.

For each activation function, 100 training iterations were performed. As a result of each iteration, the losses on the test dataset were recorded for each of the 100 epochs. The average losses for each activation function were calculated as the average of the losses for each epoch over all 100 iterations [2].

The results of the experiment are presented in the form of a graph (Fig. 5.5), which shows the average test losses for each activation function over 100 epochs of training.

The ReLU activation function showed the best performance, achieving the lowest average loss over 100 training epochs. The use of ReLU contributed to a faster and more stable error reduction compared to the other options considered, since ReLU, unlike Sigmoid and Tanh, avoids the effect of rapid saturation and provides a constant derivative for positive input values. In general, ReLU proved to be the most appropriate for the task of calibration and compensation of destabilising factors, which is consistent with its widespread use in many neural architectures focused on accurate regression estimates.

5.3 Criteria for the Influence of Hyperparameters on the Neural Network Performance

Hyperparameters, such as the size of network layers, learning rate, number of epochs, and the patience parameter for early stopping, affect the network's ability to learn and generalise. Incorrect choice of these parameters can lead to overtraining or undertraining, which will negatively affect the quality of the models. We will consider how the choice of layer size, learning rate, number of epochs, and patience parameter affects the network performance and justify the feasibility of the methods used.

Learning rate. The learning rate (α) is one of the key hyperparameters that determines the step size when updating the neural network weights during training. Formally, the update of the weight w at stage $t + 1$ can be described by the equation:

$$w_{t+1} = w_t - \alpha \cdot \nabla L(w_t), \tag{5.5}$$

where $L(w)$ is the loss function, and $\nabla L(wt)$ is the gradient of the loss function by weight w at stage t.

Graphically, the effect of the learning rate can be demonstrated on the convergence graph of the loss function during training (Fig. 5.6), if the learning rate (α) is too high, the model may "jump" the optimum and not converge. On the contrary, too low a learning rate can lead to slow convergence, which significantly increases the learning time.

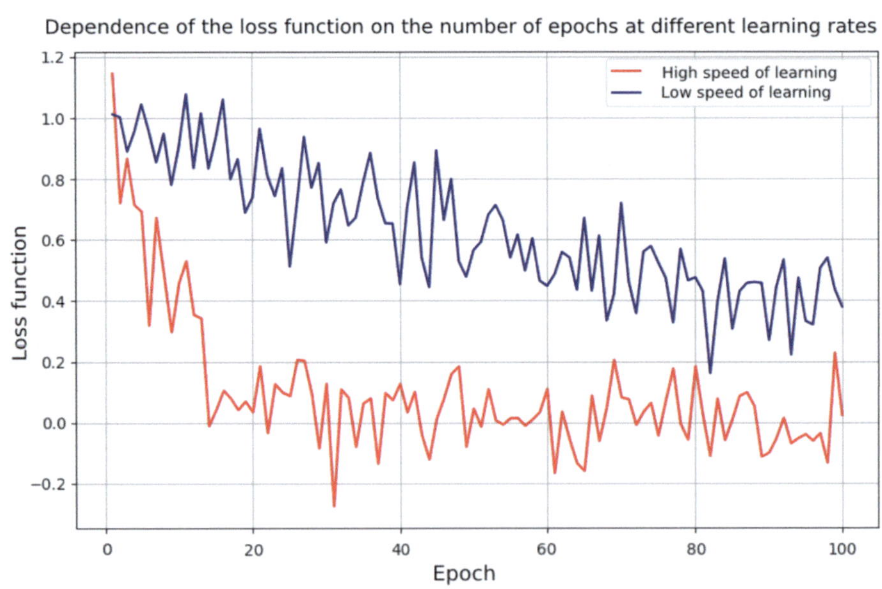

Fig. 5.6 Dependence of the loss function on the number of epochs at different values of the learning rate

Number of epochs. The number of epochs (n) determines how many times the entire training set is used to update the model weights. The training loss function changes from epoch to epoch, and usually there is a rapid decrease in losses at the beginning of training, followed by a gradual slowdown. This can be expressed as follows:

$$L(w_n) \approx L(w_0) \sum_{i=1}^{n} \alpha \cdot \nabla L(w_i), \tag{5.6}$$

where w_0—the initial values of the weights.

Figure 5.7 shows the typical change in the loss function for the training and validation datasets as a function of the number of epochs.

Too many epochs can lead to overfitting, when the loss function on the training set continues to decrease while the validation set begins to increase.

Patience parameter for early stopping. Early stopping is a method used to prevent overfitting by stopping training if the loss function on the validation set does not improve for a certain number of epochs, defined by the patience parameter. Let p be the value of the patience parameter, then the early stopping condition can be expressed as follows:

If $L_{val}(w_n) \geq L_{val}(w_{n-p})$, then stop training where $L_{val}(w_n)$ is the loss function on the validation set for epoch n.

Figure 5.8 shows how the early stopping method avoids overfitting by stopping training before the loss function on the validation set starts to grow.

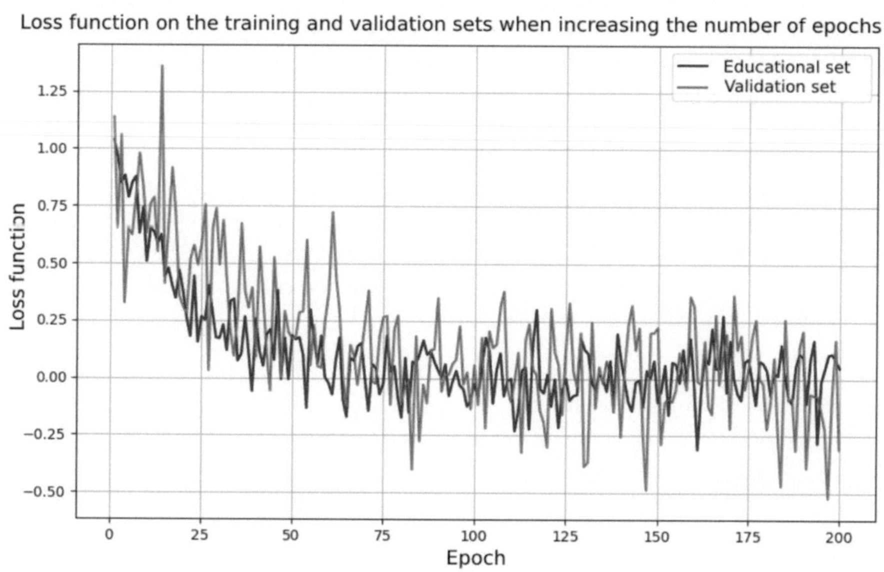

Fig. 5.7 Changes in the loss function on the training and validation sets as the number of epochs increases

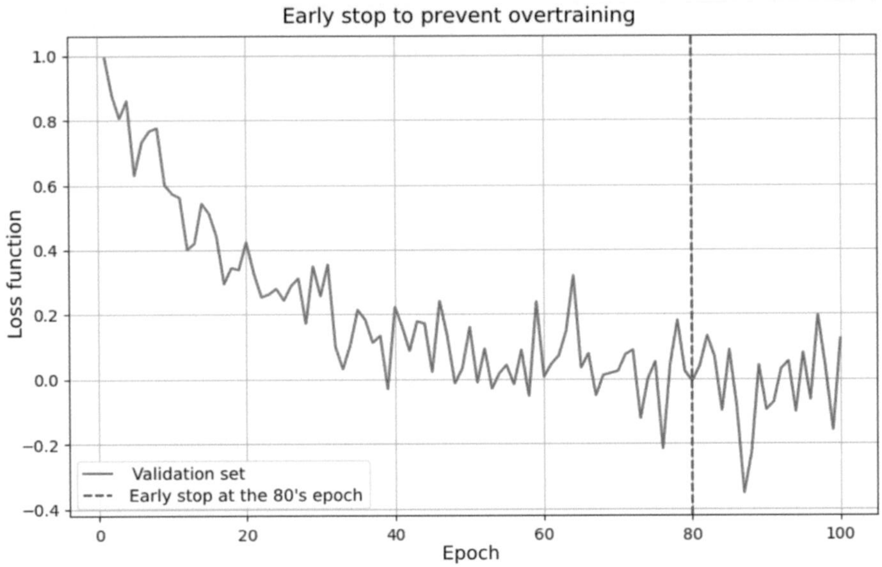

Fig. 5.8 Graph of the change in the loss function on the validation dataset

To substantiate the impact of changing hyperparameters on the performance of the NN, we selected typical key hyperparameters that affect the model performance in terms of stability and uncertainty reduction (Table 5.1).

From the selected typical key hyperparameters, 432 possible combinations were formed. Thus, 432 NN models were built and test-trained on the coordinates of the sphere surface, followed by processing of the noisy coordinates by a neural network, and obtaining the predicted coordinates of 10 test points, followed by comparing the predicted coordinates with the reference coordinates and confidence intervals. Relevant metrics are also selected to determine how changes in hyperparameters affect the measurement results. Train Loss—Losses on training data. Validation Loss—loss on validation data. Mean Square Error (MSE) of the predicted coordinates of the control points. Reduction of uncertainty before and after the coordinates are processed by the neural network. Correction rate, which indicates the percentage of model predictions that required correction after processing.

Table 5.1 Typical key hyperparameters

Hyperparameters	Typical values			
Number of neurons	32	64	128	256
Speed of learning	0.001	0.01	0.1	
Number of epochs	50	100	200	
Size of mini-packs	16	32	64	128
Regularisation	0.0001	0.001	0.01	

Table 5.2 Parameters of model No. 127

Model No.	127
Number of neurons	64
Learning speed	0.001
Number of learning epochs	100
The size of mini-batches	64
Regularization	0.0001
Train loss	0.027
Validation loss	0.026
Uncertainty X before processing (mm)	8.91
Uncertainty X after processing (mm)	8.38
Uncertainty Y before processing (mm)	8.76
Uncertainty Y after processing (mm)	8.02
Uncertainty Z before processing (mm)	8.2
Uncertainty Z after processing (mm)	7.95
MSE	0.009
Frequency of correction	0

Models with a zero-correction rate were chosen as stable. This means that all the predicted coordinates after processing by the neural network fell within the confidence interval, and no corrections were required [3–5].

Among the models with a zero-correction rate, the main criterion for selecting the most accurate models was uncertainty reduction. Based on the comparative analysis, model No. 127 (Table 5.2) was selected, which has a zero correction rate, which means that the forecasts are stable (all predicted coordinates fell within the confidence interval) and which showed the greatest reduction in uncertainty, indicating that it is able to provide the highest accuracy.

Thus, it has been experimentally determined that the best model performance in terms of stability and uncertainty reduction is provided by the following hyperparameters: number of neurons—64; learning rate—0.001; number of epochs—100; mini-packet size—64; regularisation—0.001.

5.4 Experimental Evaluation of the Effectiveness of a SIS with an Integrated Neural Network

To solve the problem of correcting non-kinematic errors, a fully connected neural network with backpropagation of the error was chosen (Fig. 5.9).

To build and train the neural network, an experimental simulation of the TDF calibration process was carried out. A synthetic data set consisting of 1000 three-dimensional coordinates was used for the experiment. The data were generated on

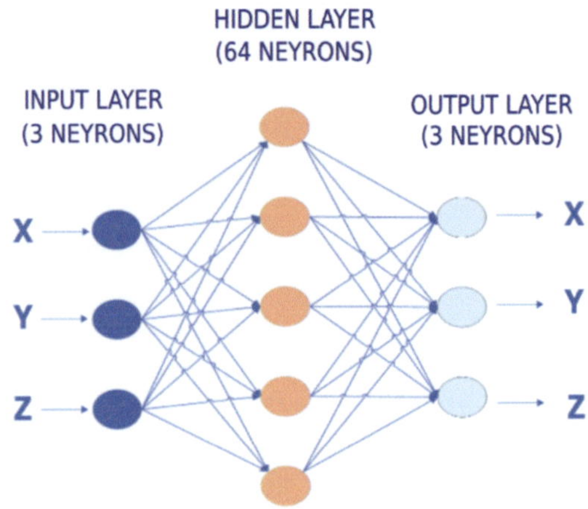

Fig. 5.9 Neural network with back-propagation of error

the surface of a sphere with a diameter of 30 mm. The data was generated using a uniform distribution. To simulate the influence of destabilising factors that may occur during measurements in real production conditions, 500 coordinates were added to the sphere coordinates according to the Gaussian distribution with a variance of 0–1.5 mm. From the 1500 coordinates obtained, 9 test coordinates were randomly selected and 1 coordinate was added to simulate an anomalous emission. The resulting coordinates (x, y, z) of the 10 test points were used as input to the neural network.

The backpropagation neural network is implemented as the Backpropagation NN class, built on the basis of the PyTorch library. The network hyperparameters are determined experimentally, as described in Sect. 4.10. The neural network consists of three main components: an input layer, one hidden layer, and an output layer.

The neural network training problem formulated as follows:

Let $p_i = (x_i, y_i, z_i)$ denote the coordinates of each of the 10 representative points obtained from the noisy data. The neural network should learn to correct errors and predict corrective corrections (Δx, Δy, Δz) to restore the coordinates that lie on the surface of the sphere [4].

The neural network implements the function:

$$f_\theta(p_i) = (\Delta x_i, \Delta y_i, \Delta z_i), \tag{5.7}$$

where θ is a set of model weighting coefficients, (Δx_i, $\Delta \mathrm{y}_i$, $\Delta \mathrm{z}_i$) is a correction vector.

The model is trained using a loss function consisting of a main criterion and a regularisation factor. The main criterion of the loss function is defined as the deviation of the quadratic sum of the corrected coordinates from the square of the ideal radius:

$$L(\theta) = \frac{1}{10}\sum_{i=1}^{10}\left[(x_i + \Delta x_i)^2 + (y_i + \Delta y_i)^2 + (z_i + \Delta z_i)^2 - R^2\right]^2. \tag{5.8}$$

This function causes the network to adjust the points so that they are closer to the surface of the sphere [5–7].

To ensure smoothness of correction, a regularisation factor is added to reduce sudden changes in the correctional adjustments:

$$L_{smooth} = \frac{1}{10}\sum_{i=1}^{10} \nabla f_\theta(p_i)^2. \tag{5.9}$$

The final loss function is then written as:

$$L_{total}(\theta) = L(\theta) + \lambda L_{smooth}, \tag{5.10}$$

where λ is the regularisation weighting factor.

The architecture of the model includes an input layer of 3 neurons that accepts 3 coordinates, one hidden layer of 64 neurons with an activation function ReLU, and an output layer with 3 neurons that generates corrective corrections with linear activation.

Forward propagation is the process by which input data passes through the layers of a neural network to become output signals. Forward propagation begins with a linear transformation operation that is performed for each layer of the network.

Let's look at this process in more detail. Let $a^{(0)}$ be the input vector of the network (i.e., the initial data). For each layer of l (where $l = 1, 2, \ldots, L$ and L are the number of layers), the following linear transformation is performed:

$$z^{(l)} = W^{(l)}\, a^{(l-1)} + b^{(l)}, \tag{5.11}$$

where $W^{(l)}$ is the weight matrix l of the layer, the size of which depends on the number of neurons in the previous layer $a^{(l-1)}$ and the current layer $z^{(l)} a^{(l-1)}$ is the activation vector (or output) of the previous layer $b^{(l)}$ is the displacement vector l of the layer, which has a dimension corresponding to the number of neurons in this layer [6–9].

The resulting vector $z^{(l)}$ is the sum of the weighted input values with the addition of offsets. It is this vector that serves as an argument for the activation function, which adds nonlinearity to the modelling process.

After calculating $z^{(l)}$, the activation function σ is applied, which determines the initial values of this layer:

$$a^{(l)} = \sigma\left(z^{(l)}\right). \tag{5.12}$$

This means that each element of the vector $z^{(l)}$ is transformed using the ReLU activation function, then for each element of $z^{(l)}$ we have:

$$a_i^{(l)} = \max\left(0, z_i^{(l)}\right). \tag{5.13}$$

Thus, the overall process of direct distribution involves the sequential application of the following formulas:

$$z^{(l)} = W^{(l)}\, a^{(l-1)} + b^{(l)};$$
$$a^{(l)} = \sigma\left(z^{(l)}\right).$$

This process is repeated for each layer, starting with $l = 1$ (where $a^{(0)} = x$ is the input) and ending with the last layer L. As a result, the output vector $a^{(l)}$ is the final prediction or response of the network [10–14].

During the back propagation of the network, the gradient of the loss function g_t with respect to each model parameter is first calculated:

$$g_t = \frac{\partial L_{total}}{\partial \theta_t} t. \tag{5.14}$$

This gradient shows the direction and rate of change of the loss function when the parameter θ is slightly changed. To update the parameters, the Adam algorithm is applied, which combines the advantages of the moment and adaptive gradient scaling methods. The Adam algorithm is described in more detail in the following steps:

First-order moment calculation (exponential smoothing of gradients):

$$m_t = \beta_1\, m_{t-1} + (1 - \beta_1)\, g_t, \tag{5.15}$$

where β_1 is usually taken as 0.9.

Second-order moment calculation (exponential smoothing of gradient squares):

$$v_t = \beta_2\, v_{t-1} + (1 - \beta_2)\, g_t^2, \tag{5.16}$$

where β_2 is usually set to 0.999.

Offset correction for the moments (since in the initial steps m_t and v_t may be biased due to zero initial values):

$$\hat{m}_t = \frac{m_t}{1 - \beta_1^t}, \quad \hat{v}_t = \frac{v_t}{1 - \beta_2^t}. \tag{5.17}$$

The parameters are updated using the formula:

$$\theta_{t+1} = \theta_t - \eta\, \frac{\hat{m}_t}{\sqrt{\hat{v}_t} + \epsilon}, \tag{5.18}$$

where η is the learning rate and ϵ is a small constant (e.g. 10^{-8}) to prevent division by zero.

Thus, instead of a simple gradient descent, where parameters are updated as:

$$\theta_{t+1} = \theta_t - \eta \frac{\partial L_{total}}{\partial \theta_t}. \tag{5.19}$$

The Adam algorithm adapts updates for each parameter separately, using historical information about the first and second moments of the gradients. This allows for effective management of parameter changes even in the case of large or noisy gradients, which significantly contributes to faster and more stable convergence when training a neural network.

To improve convergence, the strategies of reducing the learning rate and early stopping are used with the parameter patience = 20. This means that training was stopped if there was no improvement in the loss function on the validation data within 20 epochs. The model was trained on 80% of the total generated data, while 20% of the data was used for validation. After the training phase is completed, the pre-trained model f_θ is applied to the representative points selected by $K = 10$. The corrected coordinates are calculated for each point:

$$p_i' = p_i + f_\theta(p_i) = (x_i + \Delta x_i, \mathrm{y}_i + \Delta \mathrm{y}_i, \mathrm{z}_i + \Delta \mathrm{z}_i), \; i = 1, \ldots, 10. \tag{5.20}$$

These adjusted points are expected to satisfy the condition:

$$(x_i + \Delta x_i)^2 + (\mathrm{y}_i + \Delta \mathrm{y}_i)^2 + (\mathrm{z}_i + \Delta \mathrm{z}_i)^2 \approx R^2. \tag{5.21}$$

After training the model, the uncertainty of the corrected coordinates is estimated. For this purpose, confidence intervals are used:

$$CI = t \cdot \frac{\sigma}{\sqrt{n}}, \tag{5.22}$$

where t is the value of the quantile of the Student's distribution, σ is the standard deviation of the predicted coordinates, and n is the number of points [15–17].

The standard deviation of the predicted coordinates is calculated by the formula:

$$\sigma = \sqrt{\frac{1}{N}\sum_{i=1}^{N}\left(\sqrt{(x_i + \hat{x}_i)^2 + (y_i + \hat{y}_i)^2 + (z_i + \hat{z}_i)^2}\right)^2}, \tag{5.23}$$

where x_i, y_i, z_i are the true coordinates $\hat{x}_i$, $\hat{y}_i$, $\hat{z}_i$ are the predicted coordinates, and N is the number of points in the dataset [18–20].

Confidence intervals with a confidence level of 95% were calculated for each coordinate of the measured points. The real and predicted coordinates were compared according to the criterion of finding the predicted coordinates within the confidence interval (Fig. 5.10).

After training, the model was evaluated on a test dataset that was not used during training and validation. As a result, the coordinates of 10 test points were obtained before and after NN processing (Table 5.3).

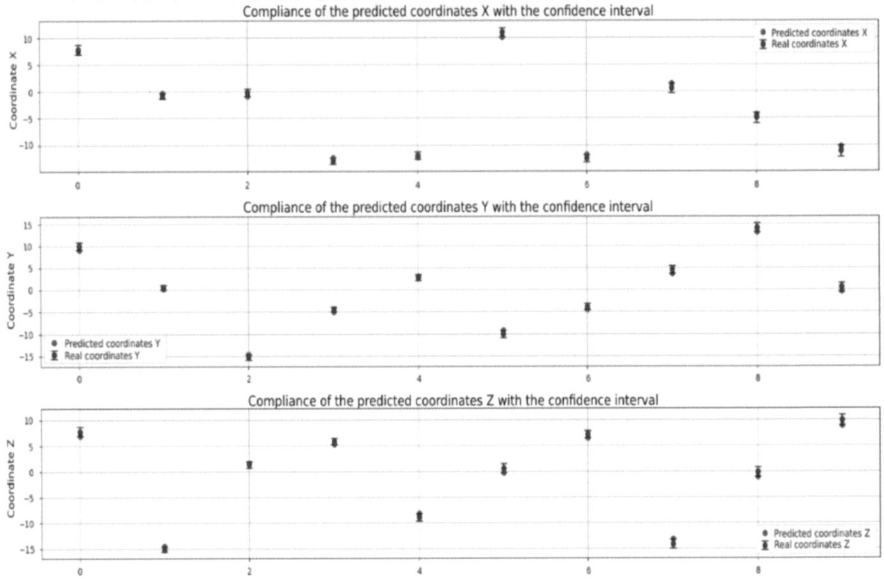

Fig. 5.10 Correspondence of the predicted coordinates to the confidence interval

Determine the distance of the point from the centre of the reference sphere (r):

$$r = \sqrt{x^2 + y^2 + z^2}, \tag{5.24}$$

where x, y, z is the coordinates of the point, r is the distance from the point to the centre.

If the diameter of the sphere is $d = 30$ mm, then:

$$R = \frac{d}{2} = 15 \text{ mm}.$$

Deviation of a point from the surface of a reference sphere:

$$\Delta r = |r - R|, \tag{5.25}$$

where r is the calculated distance of the point, $R = 15$ mm.

Table 5.4 shows the results of calculating the distance of the test points to the surface of the sphere, before and after processing the NN coordinates. If $\Delta r_{\text{after}} < \Delta r_{\text{before}}$, then the point has approached the surface of the sphere (+), otherwise—has moved away (−).

After processing with the neural network, the coordinates for most of the predicted points (9 out of 10) became closer to the reference surface of the sphere (Fig. 5.11).

Table 5.3 Coordinates of test points before and after NN treatment

Point	X to be machined (mm)	X after machining (mm)	Y to be machined (mm)	Y after machining (mm)	Z to be machined (mm)	Z after machining (mm)	Distance to the surface of the sphere before processing	Distance to the surface of the sphere after processing
1	4.09	5.11	−13.32	−13.98	−1.02	0.15	1.03	0.12
2	1.38	1.36	0.56	1.31	13.96	14.7	0.96	0.18
3	−4.4	−3.36	−12.95	−14	3.59	4.63	0.86	−0.12
4	5.5	5.93	−9.03	−8.89	−10.19	−10.62	0.32	−0.07
5	5.68	6.03	−5.42	−7.15	9.92	11.66	2.35	0.05
6	−2.3	−1.99	9.55	9.24	−12.09	−11.98	−0.59	−0.26
7	0.98	0.83	−13.64	−14.09	−5.93	−5.49	0.09	−0.15
8	11.68	11.43	7.66	8.97	3.53	2.21	0.59	0.30
9	−5.94	−6.31	7.24	6.86	−11.4	−11.69	0.25	0.05
10	−5.26	−5.05	−7.6	−8.14	10.51	11.46	1.00	0.06

Table 5.4 Deviation from the sphere surface before and after NN treatment

Point	r_{before} (mm)	r_{after} (mm)	Deviation before machining, Δr_{before} (mm)	Deviation after machining, Δr_{after} (mm)	Result
1	13.97	14.89	1.03	0.11	+
2	14.04	14.82	0.96	0.18	+
3	14.14	15.12	0.86	0.12	+
4	14.68	15.07	0.32	0.07	+
5	12.65	14.95	2.35	0.05	+
6	15.58	15.26	0.58	0.26	+
7	14.91	15.14	0.09	0.14	–
8	14.41	14.70	0.59	0.30	+
9	14.75	14.95	0.25	0.05	+
10	14.00	14.94	1.00	0.06	+

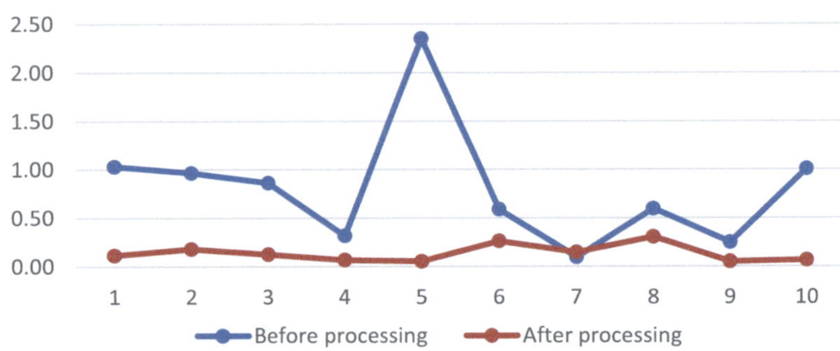

Fig. 5.11 Distance to the surface of the sphere

Thus, the use of the NN calibration procedure as an additional step, after compensating for kinematic errors using classical methods, showed a reduction in the residual non-kinematic error from 0.8 mm to 0.14 mm [21–23]. The application of this procedure allows to obtain a set of corrected measurements, for which the calculation of uncertainty type A is determined by the formula:

$$u = \frac{s}{\sqrt{n}}. \tag{5.26}$$

where s is the sample standard deviation, demonstrates significantly lower values compared to the original data. This indicates an increase in the accuracy and reproducibility of measurements in real production conditions. If the kinematic error compensation was effective, then the sample standard deviation s calculated from the corrected measurements will mainly reflect the residual non-kinematic error. However, it is worth noting that s is a statistical measure that may include small residual kinematic components if the compensation was not absolutely perfect [24, 25].

The integration of NNs allows adaptive consideration of the influence of the conditions of the real production environment, which provides more accurate compensation of residual non-kinematic errors and, as a result, a high level of accuracy of the system endpoint positioning. Thus, the additional calibration procedure using NN is an effective means to improve the stability and reproducibility of measurements, which is important for the use of IMS in production conditions. Thus, on the basis of experimental studies, the effectiveness of using a neural network to improve the accuracy of measurements of geometric parameters of power equipment parts has been confirmed. The proposed SIS integrated into the IMS allows for effective correction of residual non-kinematic errors.

5.5 Conclusions

Experimental studies were conducted to assess the effectiveness of the proposed SIS with an integrated neural network.

The architecture of a neural network for compensation of non-kinematic measurement errors is developed, which allows to increase the accuracy of measurements of geometric parameters of power equipment parts.

The influence of the architecture and training parameters of the neural network on the measurement accuracy and data processing speed is investigated.

Comparison of the results before and after neural network processing showed a significant reduction in the residual non-kinematic error by 6 times.

The proposed SIS with an integrated neural network has demonstrated the ability to effectively correct residual non-kinematic measurement errors, which allows to improve the accuracy of measurements of geometric parameters of power equipment parts.

References

1. Elatta, A.Y., Gen, L.P., Zhi, F.L., Daoyuan, Y., Fei, L.: An overview of robot calibration. Inf. Technol. J. **3**, 74–78 (2004). https://doi.org/10.3923/itj.2004.74.78
2. Piratelli-Filho, A., Araújo, J.A.: Reverse engineering of hydraulic turbine runners using coordinate measuring arms and NURBS modeling. Rev. Tecnol., Fortaleza. **30**, 114–122 (2009) https://ojs.unifor.br/tec/article/view/5282
3. Piratelli-Filho, A., Di Giacomo, B.: Uncertainty Evaluation in Small Angle Calibration Using ISO GUM Approach and Monte Carlo Method. XVII IMEKO World Congress (2003). https://www.imeko.org/publications/wc-2003/PWC-2003-TC8-003.pdf
4. Kovac, I., Frank, A.: Testing and calibration of coordinate measuring arms. Precis. Eng. **25**(2) (2001). https://doi.org/10.1016/S0141-6359(00)00057-X
5. Hyun, J.J., Hyun, K.S., Keun, Y.: Calibration of geometric and non-geometric errors of an industrial robot. Robotica. **19**, 311–321 (2001). https://doi.org/10.1017/S0263574700002976

6. Shaw, L., Weckenmann, A.: Automatic registration method for hybrid optical coordinate measuring technology. CIRP Ann. Manuf. Technol. **60**, 539–542 (2011). https://doi.org/10.1016/j.cirp.2011.03.055
7. Wilhelm, R.G., Hocken, R., Schwenke, H.: Task specific uncertainty in coordinate measurement. CIRP. (2001). https://doi.org/10.1016/S0007-8506(07)62995-3
8. Schwenke, H., Knapp, W., Haitjema, H., Weckenmann, A., Schmitt, R., Delbressine, F.: Geometric error measurement and compensation of machines—an update. CIRP. **57**, 660–675 (2008). https://doi.org/10.1016/j.cirp.2008.09.008
9. Overview of ASME B89 standards with an emphasis on B89.4.22—methods for performance evaluation of articulated arm coordinate measuring machines. NCSLI Measure. **2**(4) (2007). https://doi.org/10.1080/19315775.2007.11721401
10. Guide to the expression of uncertainty in measurement—part 6: developing and using measurement models. JCGM (2020). https://www.bipm.org/documents/20126/2071204/JCGM_GUM_6_2020.pdf
11. An investigation of measurement uncertainty of coordinate measuring machines (CMMs) by comparative analysis. J. Technol. Stud. **42**(1) (2016). https://doi.org/10.21061/jots.v42i1.a.5
12. ADDICOUNT: APCI/CPCI 1710, Synchronous Serial Interface, Technical Documentation, 2nd edn. Ottersweier, ADD-IDATA GmbH Mess und Steuerungstechnik (1999)
13. Santolaria, J., Aguilar, J.J., Yagüe, J.A., et al.: Kinematic parameter estimation technique for calibration and repeatability improvement of articulated arm coordinate measuring machines. Precis. Eng. **32**, 251–268 (2008). https://doi.org/10.1016/j.precisioneng.2007.09.002
14. Stetsenko, I.V., Boyko, O.V.: Simulation modeling system using Petri nets. Math. Mach. Syst. Kyiv. (2009)
15. Zaitsev, D.A.: Mathematical Models of Discrete Systems: A Textbook. ONAZ named after O.S. Popov, Odesa (2004)
16. Meng, J.E., Wu, S., Lu, J., et al.: Face recognition with radial basis function (RBF) neural networks. IEEE Trans. Neural Netw. **13**, 697–710 (2002). https://doi.org/10.1109/TNN.2002.1000134
17. Seshagiri, S., Khalil, H.K.: Output feedback control of nonlinear systems using RBF neural networks. IEEE Trans. Neural Netw. **11**, 69–79 (2000). https://doi.org/10.1109/72.822511
18. Bruzzone, L., Prieto, D.F.: A technique for the selection of kernel-function parameters in RBF neural networks for classification of remote-sensing images. IEEE Trans. Geosci. Remote Sens. **37**, 1179–1184 (2002). https://doi.org/10.1109/36.752239
19. Loquercio, A., Segu, M., Scaramuzza, D.: A general framework for uncertainty estimation in deep learning. IEEE Rob. Autom. Lett. **5**(2), 3153–3160 (2020). https://doi.org/10.1109/LRA.2020.2974682
20. Mitros, J., Mac Namee, B.: On the validity of Bayesian neural networks for uncertainty estimation. ArXiv (2019). https://doi.org/10.48550/arXiv.1912.01530
21. Kacprzyk, J., Zaporozhets, A., Kataiev, D.: Development of an artificial neural network for information and measurement system for controlling the geometric dimensions of power equipment. Syst. Res. Energy. **1**, 74–84 (2025). https://doi.org/10.15407/srenergy2025.01.074
22. Ovadia, Y., Fertig, E., Ren, J., Nado, Z., Sculley, D., Nowozin, S., Dillon, J., Lakshminarayanan, B., Snoek, J.: Can you trust your model's uncertainty? Evaluating predictive uncertainty under dataset shift. Adv. Neural Inf. Proces. Syst. (2019). https://doi.org/10.48550/arXiv.1906.02530
23. Seshagiri, S., et al.: Output feedback control of nonlinear systems using RBF neural networks. IEEE Trans. Neural Netw. **11**(1) (2025). https://doi.org/10.1109/72.822511
24. Rovithakis, G.A., Christodoulou, M. A.: A robust neural adaptive control scheme. In: Proceedings on 1995 Conference Decision Control. https://doi.org/10.1109/CDC.1995.480607
25. Huang, J., Zhang, M., Ri, S., et al.: High-order disturbance observer based sliding mode control for Mobile wheeled inverted pendulum systems. IEEE Trans. Ind. Electron. (2019). https://doi.org/10.1109/TIE.2019.2903778

MIX
Papier aus verantwortungsvollen Quellen
Paper from responsible sources
FSC® C105338

If you have any concerns about our products, you can contact us on
ProductSafety@springernature.com

In case Publisher is established outside the EU, the EU authorized representative is:
Springer Nature Customer Service Center GmbH
Europaplatz 3, 69115 Heidelberg, Germany

Printed by Libri Plureos GmbH
in Hamburg, Germany